Volcanoes of the Solar System

Volcanoes of the Solar System

CHARLES FRANKEL

CAMBRIDGE
UNIVERSITY PRESS

e Press Syndicate of the University of Cambridge

The Pitt Building, Trumpington Street, Cambridge CB2 1RP

40 West 20th Street, New York, NY 10011-4211, USA

10 Stamford Road, Oakleigh, Melbourne 3166, Australia

© Cambridge University Press 1996

First published 1996

Printed in Great Britain at the University Press, Cambridge

A catalogue record for this book is available from the British Library

Library of Congress cataloguing in publication data

Frankel, Charles.

Volcanoes of the solar system / Charles Frankel.

  p.   cm.

Includes bibliographical references (p.).

ISBN 0 521 47201 6 (hardcover).–ISBN 0 521 47770 0 (pbk.)

1. Planets – Geology. 2. Lunar geology. 3. Solar System.

4. Volcanism. I. Title.

QB603. G46F73 1996

551.2 '1' 0999–dc20   95-36876 CIP

ISBN 0 521 47201 6 hardback

ISBN 0 521 47770 0 paperback

CONTENTS

INTRODUCTION

This book reviews volcanism in the Solar System, starting with the volcanic features of planet Earth – which until recently were the only ones known to man – and presenting those features discovered on other planets. The space age, with its manned and unmanned lunar and planetary missions, has offered us a new, enlarged vision of volcanism, through the discovery and study of lava fields on the Moon; giant volcanic shields on Mars; a host of varied and original volcanoes on Venus; and prodigiously active sulfur volcanoes on Io, moon of Jupiter.

Although *Volcanoes of the Solar System* is written for the layman, it explores volcanic processes in enough detail to serve as an introductory guide for students of the earth sciences. In this respect, the book is structured in a two-tiered format. Odd-numbered chapters (1, 3, 5, 7) are overviews of the Earth, Moon, Mars and Venus, and are the easiest to read. They present volcanic landscapes at first glance, and highlight the space missions that led to their discovery. Even-numbered chapters (2, 4, 6, 8) explore each planet in more detail, and attempt to draw a global picture of their volcanic regimes and histories. Chapter 9 departs from this pattern by treating the volcanism of Io in one (long) chapter, which also includes a look at cryovolcanism on other moons, and Chapter 10 concludes by comparing volcanic processes across the Solar System.

For an introduction to planetary volcanism, it is therefore possible to first read the odd-numbered chapters, and leave aside the even-numbered chapters and their more specialized notions for a second reading. One can also consult this guide 'à la carte' and compose one's own itinerary through the Solar System, skipping from planet to planet in any order. A detailed index on pages 227–32 also allows a thematic approach: the entry 'basalt', for example, refers the reader to the basalts of the Earth, Moon, Mars, Venus and Io.

The following is a brief description of the contents of each chapter, so that the reader can compose his or her own reading menu.

Chapter 1 is an introduction to volcanology, based on the Earth as a model. It attempts to brush a broad picture of key concepts in volcanology, that serves as a basis for all further reading. The chapter reviews the sources of heat within planets that are responsible for volcanism; the genesis and ascent of magma; and styles of eruptive behavior.

With this in mind, Chapter 2 explores the volcanism of planet Earth in more detail, highlighting major classes of activity: mid-ocean ridge volcanism; hot spot volcanism and continental rifts; island arcs and other forms of subduction volcanism; and resurgent calderas. The chapter ends with a discussion of the pace and cycles of volcanic activity throughout our planet's history.

Chapters 3 and 4 deal with the Moon. Chapter 3 tells of the discovery of our rocky companion, and highlights the manned landings of the Apollo program: we explore the basaltic plains of Apollo 11 and 12; the Hadley Rille lava channel (Apollo 15);

and Taurus-Littrow valley (Apollo 17), the first flight to the Moon of a trained geologist.

Chapter 4 reviews the history and volcanism of the Moon, based on the analysis of the lunar samples brought back to Earth. After a look at the geochemistry and mineralogy of lunar lavas, this chapter outlines the major igneous processes that shaped the lunar surface, and attempts to outline the regional distribution and chronology of lunar volcanism.

Chapters 5 and 6 deal with Mars. Chapter 5 presents the giant volcanoes of the red planet, including a guided tour of the mighty Olympus Mons, the tallest volcano in the Solar System. Chapter 6 treats martian volcanism in more detail, beginning with the photographic coverage and chemical data provided by the Viking landers. The chapter ends with a discussion of martian volcanism through space and time.

Chapters 7 and 8 present our neighbor and sister planet: cloud-shrouded Venus. Chapter 7 tells of the painstaking discovery and exploration of this planetary inferno, through the images and chemical data of the Soviet landing probes and the first radar sketches of the surface. Chapter 8 presents the explosion of data and knowledge brought about by the Magellan radar mission: it gives an overview of the thousands of volcanoes discovered on Venus and of a few of the models put forth to explain their puzzling features.

Chapter 9 travels to the outer Solar System and the moons of the giant planets. It tells the case of extraordinary Io – a volcanic moon kept alive by the tidal forces of Jupiter and permanently erupting lava flows and jets of sulfur. The chapter concludes with the other, less active moons surveyed by the Voyager spacecraft, where ice flows and frosty deposits tell the story of cryovolcanism.

Chapter 10 wraps up our tour of the Solar System with a look at the unsung volcanic bodies like Mercury and those asteroids which underwent magmatic differentiation. This final chapter ends with a comparison of volcanic bodies across the Solar System, and a look at major trends and governing laws in eruptions and volcano building.

The picture of planetary volcanism that emerges in this book draws on the work of hundreds of planetary geologists throughout the world. I have attempted to give as objective an overview as possible of the data gathered and of framing theories and models, but I am quite conscious of my own biases: on many occasions I oversimplified the story or chose out of several theories the one that would tell the most spectacular story. I anticipate and welcome comments for the next edition!

A brief bibliography of my sources can be found on pages 221–6. It is not meant to be a comprehensive reference list but should rather be viewed as a starting point for interested students to engage in further reading.

I am most indebted to the scholars and specialists who agreed to review my chapters. They provided me with enlightening suggestions and corrections – the errors and subjective slants that remain are mine only.

I wish to thank Dr Jacques-Marie Bardintzeff, Professor at Université d'Orsay, Paris-Sud, for his review of Chapter 1; and Dr David Clague, Scientist-in-Charge of the Hawaiian Volcano Observatory, for his review of Chapter 2 on terrestrial volcanism.

Dr Harrison Schmitt, lunar module pilot on Apollo 17 and the only geologist to visit the Moon, contributed a precious review of Chapters 3 and 4 on lunar geology. So did Dr James W. Head III, Professor at Brown University's Department of Geological Sciences. I am also indebted to Jim Head and to the Department for their kind support in assisting me in my research and helping me with illustrations.

My gratitude also goes to Dr Sean Solomon, Director of the Department of Terrestrial Magnetism at the Carnegie Institution of Washington, for reviewing Chapters 5 and 6 on Mars (and to Dr Philippe Masson of Université d'Orsay, Paris-Sud, for steering me through my first draught of those chapters). Drs Larry Crumpler and Jayne Aubele of Brown University graced me with a thorough review of my Chapter 8 on Venus.

Finally I wish to thank Dr Alfred McEwen of the US Geological Survey at Flagstaff for his enlightening criticism and review of Io in Chapter 9; and to Dr Ronald Greeley, Professor of Geology at Arizona State University for his much appreciated comments on Chapter 10.

Most of the illustrations in this book were graciously provided by learning institutions, research centers, and individual scientists and their publishers: I am indebted to all. Due credit is given for each image in the corresponding caption.

The concept of *Volcanoes of the Solar System* originated in France with the Armand Colin publishing house, under the guidance of scientific director Jean-Luc Blanc and collection supervisor Albert Ducrocq. It was a pleasure to work with the Armand Colin staff as it was to work with Cambridge University Press for this rewritten and updated English version. I hope readers will sense the passion and dedication that so many people invested in this book. Bon voyage!

1 The Earth as a model

Volcanoes in space

Volcanoes and lava fields are found on all the rocky planets of the Solar System: on the Earth and Moon, Mercury, Venus, Mars, and many smaller satellites of the outer planets.

Some volcanic provinces have been extinct for billions of years – starting with our own Moon – and constitute rocky exhibits of the early days of planetary volcanism. Other features are much younger, such as the volcanoes of Mars and Venus; and two planetary bodies are conspicuously active today: our own planet Earth, and Jupiter's extraordinary little moon Io.

The widespread occurrence of volcanic provinces – past or present – throughout the Solar System stresses the importance of volcanism in the creation and evolution of planets. This prominent role is due to the fact that planets are born hot, and continue to produce heat – albeit at a declining rate – throughout their lifetime. Magmatism and volcanism are the fundamental mechanisms that allow planets to cool off, by transporting heat upward to the surface, where it is radiated to space.

Large rocky planets produce more heat than smaller planetary bodies, and call on volcanic activity to evacuate the surplus – the prime example being the Earth. We will therefore begin our tour of the Solar System by presenting the volcanic life of our own planet. The rules and models we derive for terrestrial volcanoes will then serve as guidelines for discussing volcanic features on other planets.

The Earth as a model

On our own planet, volcanoes received little attention in the early days of geology: scientists were drawn by the neat, geometric layers of sedimentary rock and their precious fossils, and volcanic intrusions and lava flows were perceived as oddities, upsetting the natural order. Most inactive volcanoes went unrecognized for centuries, as did the striking chain of lava cones and domes of central France, which were not identified as igneous bodies until 1752, and not without some debate!

The emergence of volcanism as a major player in the earth sciences was delayed until the dawn of the twentieth century, when the discovery and study of new volcanoes – namely those of Africa and the Pacific rim – sparked the interest of geologists around the world. Even more dramatic was the realization, shortly after World War II, that the deep ocean floors – accounting for two thirds of the Earth's surface – were predominantly made up of basalt, under a thin veneer of sediment.

The revelation of this extensive underwater volcanism in the 1950s was one of the major breakthroughs that led to the concept of plate tectonics – a revolutionary, unifying model that encompasses and relates all key geological processes on Earth, including the function and distribution of earthquakes and volcanic activity.

FIG. 1.1. The Earth,
viewed by Apollo 17.
Credit: NASA.

This global model of plate tectonics is based on the understanding that the ocean floors undergo volcanic expansion: the Earth's lithosphere (its rigid outer shell) stretches and tears along rift zones, and the cracks fill up with rising magma, chilling in place to form new crust.

Along with other fault zones, these rifts split the Earth's surface into a dozen tectonic plates, which undergo differential motions with respect to each other. Some plates tear apart at spreading margins (mid-ocean ridges), increasing in area; elsewhere plates converge at destructive margins, where one plate buckles under another and plunges back into the deep mantle (subduction zones); and in yet other places, two plates might slide past each other with no gain or loss of crustal material (strike-slip faults).

The slow differential motion of lithospheric plates at the Earth's surface is the result of underlying convection currents: the hot, plastic mantle at depth is churned by temperature and density differences, rising in places to the base of the rigid lithosphere, which it stretches and infiltrates with magma (rifting). Having released a good deal of its heat in the process, the current of plastic mantle rock becomes cooler and denser, and sinks back into the lower mantle, completing its convective loop.

The churning of the mantle is slow indeed: at depth, the hot plastic rock flows at rates of 1 to 10 cm/year, and horizontal displacement rates of the rafting lithosphere are of the same magnitude. The Atlantic Ocean, for example, stretches at an average rate of 2 cm/year: it takes a century for Europe and North America to separate by a couple of meters.

At this rate, it is difficult to perceive any change in the geography or geology of the Earth's surface over the span of a human lifetime: in essence we are left to contemplate one still frame of an otherwise action-packed film. The picture is indeed quite different if we reason in millions of years, on the scale of geological time periods: at an average rate of 2 cm/year, the Atlantic Ocean took only 150 million

FIG. 1.2. Mount Etna, imaged by Landsat 1. The tallest volcano of Europe (3340 m) built its 40 km-wide shield on Sicily's sedimentary platform. White patches of snow crown the summit craters. East of the summit, the chevron feature open toward the sea is the Valle del Bove graben, scene of many lateral eruptions. The volcano's largest and most recent lava flows are visible as dark radial streaks: notice the 1669 flow striking south from the summit along the whitish vapor plume, down to the city of Catania on the coast. Credit: NASA/Landsat, image processing by SEP/Vizir.

years – 3% of geological time – to broaden from a narrow gulf to an ocean 4000 km wide. If the 4.5 billion years of Earth history were condensed into a film two hours long, the birth and growth of the Atlantic Ocean would be but a four minute scene!

Distribution of volcanoes on Earth

To get an 'instant' picture of terrestrial volcanism, let us take a look at the last frame of the geological picture – say the volcanoes in activity over the last 10 000 years, the extent of human civilization. Over this period, no less than 1343 volcanoes are known to have erupted at least once, and the historical record lists more than 5000 individual eruptions occurring over the past 2000 years.

The volcanoes of Earth are therefore numerous and highly active. They also prove to be very diverse in shape, size, and eruptive behavior. In western history, this wide gamut of volcanic morphology and activity was noticed early on around the Mediterranean basin: philosophers and scientists of the classical era observed the ash eruptions of Vesuvius, lava flows running down the slopes of Etna (see fig. 1.2), sputtering magma shooting out of the vents of Stromboli, and explosive bursts rocking Vulcano. This Mediterranean legacy is still strong: when it comes to describing volcanic behavior on the Earth, as well as on the Moon and other planets, we still talk of *plinian*, *strombolian*, and *vulcanian* activity.

Early 'showcase' volcanoes were observed to be coastal (Vesuvius, Etna), or islands in their own right (Vulcano, Stromboli). Looking for causes, Aristotle suggested that the sea played a major role in their origin and behavior: waves, he believed, broke on the threshold of marine grottos, and blew draughts to stoke subterranean fires.

FIG. 1.3. Mount Saint Helens in the Cascades chain of the northwestern United States. Subduction volcanoes of the kind are often built of volatile-rich silicic magmas: viscous emprisonment of the exsolved gases can lead to pressure build-up within the cone or dome and lead to cataclysmic eruptions. A landslide on Mount Saint Helens' northern slope triggered its explosive eruption of May 18, 1980, when the abrupt pressure release sent a pyroclastic blast across the land, followed by a plinian plume rising high into the stratosphere. In the caldera gutted open by the explosion, a dome of viscous dacite is building anew, crowned by a plume of water vapor. Credit: Photo by L. Topinka, USGS, Cascades Volcano Observatory.

From the seventeenth century onwards, with the great voyages of discovery, the roster of volcanoes began to grow: in 1650, Varenius listed 21 volcanoes; two centuries later, Humboldt accounted for 407; and that number has tripled today with the 1343 items of the Smithsonian catalog.

The proximity of most volcanoes to the sea reflects the match between volcanism and tectonic plate margins. The greatest concentration of subaerial volcanoes is the 'Circle of Fire' around the Pacific rim, stringing together the volcanic arcs of *subduction zones*: the chain runs from Indonesia and the Philippines northward through Japan, the Kuril Islands, Kamchatka, eastward through the Aleutian arc to North America, down the western coast of the USA (Oregon, Washington, and California), Central America, and the Andes of South America. Out of the 1343 recognized, active volcanoes on Earth today, two thirds (900) belong to the Circle of Fire.

Volcano catalogs do not tell the whole story, however, since they only take into account visible volcanoes – those on continents, and those emerging as islands, or erupting noticeably close to the water surface to be detectable. However, as we shall now see, the largest population of volcanoes on Earth lies hidden thousands of meters below sea level, on the deep ocean floor.

The discovery of underwater volcanoes

One of the key realizations that led to the concept of plate tectonics, as we mentioned earlier, was the discovery of the volcanic nature of the deep sea floor: ocean basins are characterized by abyssal plains of basalt; scattered volcanic shields called *seamounts*; and long chains of mid-ocean ridges snaking through the abyss.

The large-scale reconnaissance of these underwater features, in the wake of World War II, was dictated by the rising strategic importance of submarines, and the need to map out their cruising environment. As military mapping progressed in the forties and fifties, the true nature of our planet was revealed – a world dominated by large volcanic basins in a state of constant crustal renewal.

This is so much in contrast with the visible part of the Earth's surface – the continents showing relatively little volcanic activity – that we might ask ourselves whether an explorer from outer space, orbiting the Earth as we do other worlds, would be able to tell our planet's profound dichotomy. The speculative answer is yes. Space flight around the Earth shows very slight variations in orbital speed at each pass over an ocean: such gravimetric 'signatures' point to the existence of deep basins under each ocean, floored by distinctively dense rock. In other words: volcanic basins, as are also found on the Moon, Mercury, Venus, and Mars...

It should also be pointed out that the topography of the hidden ocean floor can be guessed from orbit in some detail, despite the opaque blanket of water – or rather with the help of it. Geodesic satellites are capable of measuring their elevation over the water surface by laser altimetry, with a precision of a couple of centimeters: such topographical maps of the ocean surface show subtle swells and troughs – independently of waves and tides – which betray the differential gravity pull of underwater seamounts, trenches, and ridges on the water mass.

Thus, at the close of the twentieth century, we now have the tools to map our long-ignored ocean basins: satellites for global surveys; and sonar and other craft-based instrumentation for more local, high resolution mapping. These new tools constitute a welcome breakthrough, since most maps of the ocean floor today are still the monopoly of the great naval powers, and are not readily available to the public. Those which are released are large-scale maps with coarse resolutions of 30 to 40 km; and only a few quadrants reach down to the 10 km precision level, when the area is declassified because it no longer presents a strategic interest.

In retrospect, we are faced with something of a paradox: we have a better global coverage of our sister planets than we do of the Earth itself, because of this lack of knowledge – censorship, should we say – of our own ocean basins. An average 40 km resolution for the ocean floor is a far cry from the 200 m resolution averaged on our space-probe imagery of the Moon, Mars, and Venus! But until better charts of our

FIG. 1.4. Map of volcanic seamounts on a section of the Mid-Atlantic Ridge, near 24° N (Sea Beam sonar data). The population density is roughly 80 seamounts per 1000 km² along the axial rift zone, with the flat-topped cones averaging 50 to 100 m in height and several hundred meters in diameter. Seamounts are shown with summit areas unshaded and slopes hachured. Straight lines with ticks mark the crests of ridges, and stippled areas represent possible flows from fissure eruptions. Credit: From D. K. Smith and J. R. Cann, 1992. The role of seamount volcanism in crustal construction at the Mid-Atlantic Ridge, *J. Geophys. Res.*, **97**, 1645–58.

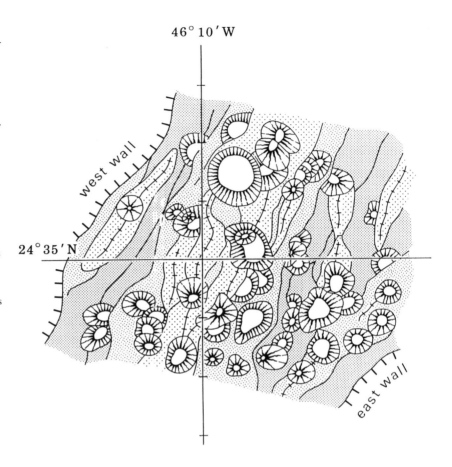

ocean floors are generated and released, we must content ourselves with a 40 km resolution for two thirds of the Earth's surface.

Such coarse maps still allow us to identify several thousand large submarine volcanoes, the size of Vesuvius. As we shall see in the next chapter, research vessels with diving robots and sonars can also map out smaller features in selected target areas. These 'spot checks' reveal that the mid-ocean ridges – and to a lesser degree the abyssal plains – are peppered with small volcanic mounds a few tens of meters in height, and several hundred meters in width (see fig. 1.4). Extrapolated to the entire ocean floor, this population might well group over a million features, hundreds of times more volcanoes than on land.

The origin of heat

Because of the widespread ocean cover, we have seen how tricky it is to map out the surface of our own planet. It is even more difficult to crack the Earth's third dimension, and picture what goes on underground. The depths of the Earth are forever barred from human experience: the crushing weight of encasing rock, as one drills down, combined with the rising temperature, limit man's physical exploration to a mere dent in the surface crust. The deepest manned mines bottom out at 2000 m, and the record for automated coring is held by the Kola drilling project in Siberia, currently bottoming at 12 000 m.

Luckily, there are other ways to probe the depths of the Earth. Remote-sensing instruments are set up at the surface – seismographs, magnetometers, and heat probes – collecting data from depth at many sites around the world. Heat probes, for instance, measure the heat flow reaching ground level – valuable data for understanding the thermal budget of the Earth, and the volcanic potential of each area.

Accretional heat

The first question we might ask is where this heat comes from, which drives plate tectonics and volcanism. Planets tend to heat up for three reasons, two having to do with the very way in which they were formed.

The Solar System owes its existence to the gravitational collapse of an interstellar cloud of gas and dust. Heat was generated in the process, through the transformation of the falling matter's potential energy into kinetic and thermal energy. The heart of the collapsing system became the Sun, overheating to the point of igniting thermonuclear fusion; and around the newborn star a disk of leftover matter tore up into discrete rings.

Such planetary rings can be thought of as orbital highways, jammed with rocky snowballs hurtling down the lanes, some gently nudging past each other, and others colliding in a most destructive way. This was a war of worlds in the truest sense, with protoplanets accreting in runaway fashion, only to break up again at the next major collision. After an initial period of chaos, one dominant body emerged in each ring, and progressively engulfed the runners-up to become that region's undisputed master. Such runaway sweep-ups were highly energetic, each new impact bringing a blast of heat to the surface of the growing body.

An accreting planet also experiences internal heating, due to the compression of the massive body under its own weight: cumulated with the energy of impacts, gravitational compression raised the temperature of the growing Earth to over 1000 °C in a matter of a few million years. This is what is meant by a planet's *accretional heat* – the first of its three heating modes.

The iron catastrophe

The second source of a planet's heat is also gravitational in origin, and has to do with the reorganization of the planet's matter into layers of contrasting density.

Initially, the cosmic dust and debris accreted by a planet is a hodgepodge of very diverse material: some is fluff of low density, rich in water, carbon, or other volatiles; while other projectiles are dense nuggets of iron and nickel. We can still see this range of source material in the make-up of meteorites reaching Earth today. Because of this heterogeneous make-up, the Earth soon found itself in a state of flux: already half-molten by the elevated temperature, the iron nuggets began to sink through the lighter material toward the planetary center, releasing potential energy in the process.

Again, this heat release is a physical consequence of matter dropping through a gravitational field, but this time the process acts inside the planet, rather than on the outside. As the hot Earth acted like a giant decanting filter – its iron and related elements sweeping downward – the temperature of the host body shot up in a dramatic way.

This sinking of the core, also known as the 'iron catastrophe', is thought to have occured roughly 500 million years into the life of the planet, when it was already hot enough for iron to melt and seep down. Estimates for the catastrophic temperature rise which resulted are as high as 2000 °C, bringing the newly-formed iron core and silicate lower mantle to temperatures in excess of 4000 °C.

Radioactive heating

In addition to accretional energy and iron sinking, the third source of a planet's heat is radioactive decay.

Among the many different atoms trapped inside a planet, some are naturally unstable: they tend to transmute into lighter, more stable elements, releasing particles and energy in the process. This radioactive heating is by far the most important and the longest-lasting source of energy within the Earth: its effect was noticeable early on, under the impulse of fast-decaying elements like aluminum-26 – fast in the sense that half the stock of Al-26 disintegrates over a time period of less than a million years on average (the element's *half-life*), bringing an early boost of energy to a newly formed planet.

Different species of atoms have different rates of decay, so that the radioactive heat release is spread out over the entire history of the Earth – although constantly declining as the radioactive fuel is expended. Potassium-40 decays relatively quickly (half-life of 1.28 billion years), and expended most of its radioactive potential in the first half of the Earth's history, whereas elements which decay more slowly like uranium-238 – an atomic species with a long half-life of 4.5 billion years – and thorium-232 (with an even longer half-life of 14 billion years), have expended only one half, and less than one quarter of their fuel stock respectively, and will keep providing more heat over the aeons to come.

The Earth stocks most of its heat within its insulating layers of rock, and transmits only a trickle to the surface where we measure it as heat flow. On average, this heat flow amounts to a tenth of a watt per square meter ($80\,mW/m^2$) at the surface: in order to light a 50 watt lightbulb, we would need to perfectly trap all the outflowing energy over an area the size of a tennis court. This might seem like a small trickle of heat, but integrated over the entire surface of the Earth, the total heat flow reaches 10^{21} joules per year, ten times the energy generated and expended by mankind over the same interval – oil, coal, gas, and nuclear combined.

Of this constant heat flow, an estimated 20% is heat left over from the early days of planetary accretion and iron core formation – only now reaching the surface – whereas

80% is supplied by radioactive decay from uranium, thorium, and potassium sprinkled throughout the planet.

Getting rid of the heat

When heat has a hard time leaving a body – and this is the case with a large, rocky planet like the Earth, which acts as a giant thermos bottle – the temperature of the host body rises.

We previously saw that accretional heat, and 'short-fused' nuclear elements, quickly brought the internal temperature of the young Earth to 1000 °C, and then up to 2000 °C within a few hundred million years. Iron then began to melt and sink to the center, releasing a 'heat flash' that brought the temperature of the deep layers to over 4000 °C. Temperatures at the center continued to rise to reach an estimated maximum of 6000 °C, and have decreased only slightly since, the amount of heat seeping out at the surface striking a balance with the continuous production of radioactive heat at depth.

This steady state is a difficult balance to achieve, since heat production and heat escape are very different processes, and don't often match up. Heat production inside a rocky planet is indeed proportional to the amount of radioactive atoms locked inside the body – i.e. to a planet's volume. Heat escape, on the other hand, is proportional to the surface a planet provides for the rising heat to radiate away – i.e. to the planet's area. These two quantities – volume and area – depart greatly from each other as we run up the gamut of planetary sizes: a small planet has a relatively large area compared to its volume, and can thus efficiently dispose of heat production and cool off. But when a planet has a radius (r) of several thousand kilometers – as is the case with the Earth – the volume has grown proportionately so much more than the area (r^3 versus r^2) that cooling is no longer in step with heat production: temperatures become so high that the planet begins to melt.

Inside the Earth

When a planet begins to melt, it does so at very specific levels, which are dictated by temperature and pressure gradients, as well as by mineral and volatile make-up. Contrary to expectations, the deeper layers of a planet are not necessarily the most molten, the reason being that the effect of high pressure – which inhibits melting – often supersedes the effect of high temperature. This is most obvious in the case of the Earth, which has only two zones where matter is liquid or semi-liquid.

Our knowledge of the inside of the Earth comes mostly from the study of earthquakes – tears and breaks of the rigid lithosphere which send shock waves ringing through the planet. The analysis of the paths followed by seismic waves – their reflections, refractions, and speed changes – allows us to locate the boundaries between different layers at depth: molten zones in particular stand out because they dramatically attenuate wave propagation. It is also possible to deduce

FIG. 1.5. Internal structure
of the Earth. Soon after
its accretion, the planet
segregated into distinct
layers of contrasting
compositions and
densities: heavy iron sank
to create the core, while
silicates were left to form
the mantle and became
capped by a thin crust.
Credit: From La
Documentation Française.
Documentation
photographique 6107,
Les volcans, 1990.

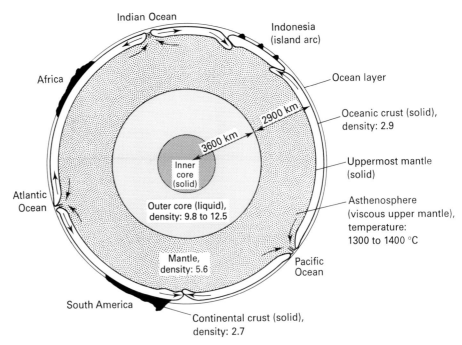

FIG. 1.5. Internal structure of the Earth. Soon after its accretion, the planet segregated into distinct layers of contrasting compositions and densities: heavy iron sank to create the core, while silicates were left to form the mantle and became capped by a thin crust. Credit: From La Documentation Française. Documentation photographique 6107, Les volcans, 1990.

the make-up of each layer by comparing the velocity of seismic signals with those of various rock compounds measured in the lab.

The model we now have of the Earth is summarized in fig. 1.5. At the center lies the inner core – a hot 'nugget' of solid iron (with traces of nickel and sulfur) at temperatures over 5000 °C. This inner core stretches from the center of the Earth (6378 km) to 5150 km from the surface.

At 5150 km, the declining pressure triggers a phase change, and the iron compound switches from solid to liquid. This outer core is the major liquid layer within the Earth: we can think of it as a 'hydraulic' ball-bearing, and it is indeed held responsible for several key processes, most importantly a dynamo effect which generates the Earth's magnetic field.

The liquid outer core comes into contact with an overlying solid, but somewhat plastic, silicate mantle 2890 km from the surface, a turbulent transition level known as the *D″ zone*. It is believed that turbulence and heat release in this D″ transition zone set up some of the flow patterns observed in the overlying mantle, such as the *hot spots* discussed in the next section.

The solid, silicate mantle stretches from the D″ boundary practically up to the surface, and is divided into several sub-zones as a function of pressure and pressure-dependent mineral assemblages. The lower part of the mantle is made up of high-pressure minerals known as perovskite, magnesiowüstite, and stishovite. Their constituent atoms are mainly silicon and oxygen, magnesium, iron, calcium, and lesser amounts of other atoms. As we rise closer to the surface, distinct pressure thresholds are crossed (a major one lies at 670 km depth), and the crystals relax to less dense mineral forms such as pyroxene, olivine, garnet, and – higher yet – spinel, and feldspar.

Between 200 km and 100 km depth, the declining pressure causes some of the minerals to melt, forming a partially-molten layer known as the *Low Velocity Zone* – or *LVZ* for short (low velocity because seismic waves slow down when they cross this rock 'mush'). This upper section of the mantle is the source of most magma erupted at the surface, and is of great interest to volcanologists. Above it, the mantle again fully recrystallizes to form a rigid roof, topped with a crust of igneous and sedimentary rock: the *lithosphere* on which we live, broken up into a mosaic of tectonic plates.

Convection currents

Despite being kept solid by overriding pressures over most of its range, the hot silicate mantle behaves plastically. It deforms and creeps underground in large, convective loops.

Convection currents are familiar to us in everyday life: when we heat a liquid on a stove, hot currents rise in the middle (over the heat source), and after releasing some heat at the surface, sink back along the sides to charge up with more heat at the bottom. Hot plastic rock behaves in much the same way, but at a dignified, slow pace.

These large-scale dynamics are what drive plate tectonics, and volcanic eruptions at the surface. Indeed, when a convective loop brings hot mantle material near the surface, the pressure drop triggers melting among the minerals, and droplets of magma ooze forth. Since the melt has a lower density than the minerals it replaces, the buoyancy of the mixture is enhanced, making it rise ever more rapidly. This runaway cycle brings magma very close to the surface, and triggers abundant volcanism, most notably under the mid-ocean ridges.

Convective loops of this sort are thought by many to churn only the upper mantle. But hot and buoyant material can also originate much deeper in the mantle, as far down as the mantle/core boundary. Caused by thermal anomalies and other instabilities at depth, these upwellings rise vertically as straight plumes, rather than convective loops. Named *hot spots*, they fuel a distinctive type of volcanism at the surface.

Besides 'turnover' melting below ridges, and deep-seated hot spots, a third type of melting environment exists within the Earth: *subduction zones*. Where old and dense lithosphere bends and sinks back into the mantle, magma is generated along the descending slab, as it heats up and flushes with volatiles.

Magma genesis

The complexity of terrestrial volcanism is enhanced by compositional factors. Like the crust on which we live, the mantle is not a homogeneous, well mixed batter. Not only does it vary in its gross mineral make-up as a function of pressure and depth, it also shows more subtle chemical variations, vertically and laterally across the Earth. Indeed we can best picture the mantle as resembling a giant marble cake.

Occasionally we get access to chunks of mantle brought up to the surface, which allow us to check on compositions. Some slabs of upper mantle rock appear in mountain folds and other dynamic settings, where tectonic accidents have thrust slices up into full view. These slabs are called *ophiolites*: they are somewhat eroded and mangled, but provide us with invaluable information on the mineral and chemical make-up of the upper mantle. As predicted by seismic models, the greenish mantle rock exposed at the surface is composed principally of crystals of olivine, pyroxene, and spinel. The assemblage is named *lherzolite*, or – in its altered, lustrous form – *serpentine*.

On other occasions, we get to sample mantle rocks from even lower levels, when deep-rooted volcanic eruptions cough up pristine chunks of unaltered material to the surface, known as mantle enclaves or *xenoliths*.

These samples confirm that the mantle displays slight chemical variations from region to region. But superimposed on these fine variations in the source region, there is a second process that has a major influence on the make-up of any magma rising to the surface: the proportion of mantle rock that melts.

Indeed, when a surge of temperature – or a release of pressure – starts melting the mantle, it is not the entire rock body that melts uniformly, but only a fraction. The various minerals that make up the mantle have different melting temperatures, and the most fusible crystals will liquefy first. As the temperature keeps rising – or the pressure keeps dropping – more refractory minerals will break down in turn and join the melt.

The chemical composition of 'freshly pressed' magma is thus highly dependent on the fraction of mantle rock that melts. If the fraction is low, the resulting trickle of magma will be enriched in fusible elements like sodium and potassium; if the fraction is high, late-melting, refractory elements such as calcium and magnesium will enter the melt and alter its overall composition. This gradual process, called *partial melting*, weighs heavily in controlling the composition of any magma in the making.

The ascent of magma

Once it oozes out of its rocky matrix – along fine fractures in the rock – magma collects in larger fissures and starts ascending to the surface. This upward motion is a natural response to Archimides' law: liquid magma is less dense than the rock matrix from which it derives, and it will rise by buoyancy through any cracks it can find or create along the way. The greater the density difference between the magma and the encasing rock, the greater its ascent velocity.

In order to reach the surface, a magma needs to rise pretty fast – in the range of one or two meters per second – because if the rise is too slow, the magma has time to cool along the way through contact with the encasing rock. Slow-rising magma is therefore at risk of solidifying and coming to a halt. These underground lenses of solidified magma that never make it to the surface are called *plutonic* rocks, in contrast to *volcanic* rocks at the surface. Where they infiltrate horizontal and vertical

FIG. 1.6. Development of a shallow magma reservoir in a volcanic edifice. One of the main parameters that rules the ascent of magma is its difference in density relative to the encasing rock. Hot magma rises by buoyancy as long as it is less dense than its surroundings. When this is no longer the case (because the wall rock it reaches is riddled with vesicles, fractures, etc.), the magma stalls in this 'neutral buoyancy zone', creating a magma chamber and attempting to propagate sideways. Credit: Adapted from J. W. Head and L. Wilson, 1992, Magma reservoirs and neutral buoyancy zones [. . .], *J. Geophys. Res.*, **97**, 3877–3903.

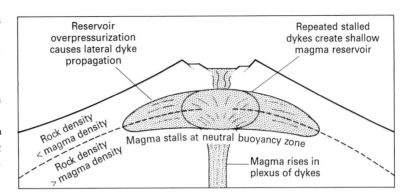

cracks before chilling in place, these magma bodies are respectively called *sills* and *dykes*.

A good way for magma to reach the surface is for partial melting to affect a large fraction of the source region: large quantities of melt will indeed stay hot longer than smaller amounts (in the same way a bathtub of hot water cools more slowly than a mugful), and will experience proportionately less cooling on the way to the surface. Note that the tectonic setting is also important in this respect: in distensional zones of the crust, large faults let more magma seep up – and oppose proportionately less cooling area and less viscous drag to the flow – than smaller fractures in 'tighter' tectonic zones.

The ascent of magma is therefore quite a complex matter. Some melts never make it to the surface; others rise directly to the surface without stopping on the way – as seems to be the case with Mount Etna for instance – and some batches ascend in a stepwise fashion, stagnating at various levels before changing conditions – a surge in temperature, a drop in pressure, or the wrenching open of new cracks – allow the magma to resume its ascent.

Magma chambers

The stagnation of magma in underground 'pockets' is a common occurrence under many volcanoes, where such reservoirs are called *magma chambers*.

Magma chambers represent the levels at which the rising magma reaches a temporary equilibrium, having cooled to the point of matching the density of the encasing rock (see fig. 1.6). No lift power is left and the magma spreads laterally, infiltrating cracks and melting down whole panels of host rock.

Magma chambers can be visualized under active volcanoes by means of seismic tomography: melt pockets show up on three-dimensional charts, since their fused matter slows down seismic waves. A good example is the magma chamber under Kilauea volcano in Hawaii. There, the chamber is imaged as a lens 4 km thick, stretching 2 to 6 km below ground level. It is fueled by magma piped up from a source region at 60 km depth, and acts as a giant manifold to distribute the molten rock laterally to the rift zones. Presently (in 1995), it is the eastern rift that gets

FIG. 1.7. The snow-capped summit of Kilimanjaro volcano, Africa's highest peak (4930 m). Built over one of Africa's many hot spots along the East African Rift. Kilimanjaro has not erupted in historical times. Its summit is collapsed in a series of concentric calderas, due to multiple eruptions and withdrawals of magma from under the summit's 'roof'. Credit: Photo by J. Besseville, LAVE.

supplied, and the magma chamber has the shape of an elongated gourd, its neck stretching below Kilauea's eastern chain of craters.

Near-surface magma chambers also betray their position and shape when they purge out during massive eruptions, or when magma is drawn back down the 'pipes', leaving the chamber 'honeycomb' with no mechanical resistance, and causing the collapse of the roof: the depression propagates upward and the caved-in structure at the surface is called a *caldera* – cauldron in Portugese. Calderas are typical of volcanoes with shallow magma chambers (see fig. 1.7).

Most magma chambers are encountered some 10 to 30 km below ground level: magma can stay stored in such pockets for centuries, even millennia, before cooling to stone, or growing new shoots upward if conditions allow. At Saint Vincent's Soufrière in the Caribbean, there are two magma chambers located under the volcano: a deep one at the 30 km level, with a pipe leading up to an upper chamber nested 10 km from the surface.

Chambers tend to regulate the pace of eruptions, since they store and distribute the rising magma. They also play a chemical role, because in times of storage, magma cools and crystals form. The first-born refractory minerals settle out of the liquid, and withdraw certain atoms from the melt preferentially to others, changing the chemical proportions in the remaining liquid.

This chemical evolution of magma, known as *crystal fractionation* or *differentiation*, is a complex operation. Denser crystals sink to the bottom of magma batches, where they constitute a natural 'slag' (chromite crystals segregate in this way). Lower

FIG. 1.8. Differentiation trends in magma appear most distinctly when the consecutive lavas of an eruption are plotted in a $Na_2O + K_2O/SiO_2$ diagram. Arrows show the chemical trends followed by tholeiitic basalt (lower curve), andesitic basalt (middle curve), and alkali basalt (upper curve) as differentiation proceeds. The lower two suites evolve towards dacite, whereas alkali basalts evolve towards either phonolite or trachyte. Rhyolite can be an extreme case of differentiation, but is often due to the contamination of the magma by siliceous crust. Credit: From Jacques-Marie Bardintzeff, 1993, *Volcans*, Armand Colin, Paris.

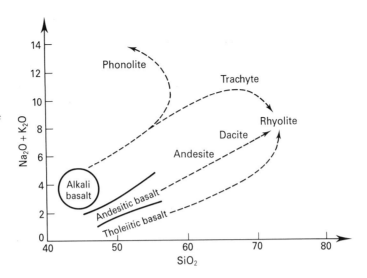

density minerals, such as plagioclase feldspar, migrate instead to the top and sides of the chamber. More levels of complexity are added by the circulation of convection cells within the chamber; purges of magma out of the chamber during eruptions; new injections of fresh magma from depth; and thermal erosion and 'digestion' of wall rock on the periphery of the chamber.

In many cases, distinct chemical trends can be followed from one eruption to the next, and the succession of changing lavas is known as a *lava series* or *suite* (see fig. 1.8).

The *calc-alkaline suite* is the most common differentiation trend on Earth, and is found both in hot spot and subduction settings. In a typical calc-alkaline suite, as exemplified by lavas from the Puy de Dôme chain in central France (see Table 1.1), the first – most pristine – lavas to erupt are basalts rich in iron (12%), magnesium (7%), and calcium (10%) oxides. As the chamber cools, refractory minerals chill out on the edges, and withdraw a great deal of these metals from the melt: the next lavas to erupt will be less ferromagnesian, and they are called leucobasalts, since they are paler. These have less iron (10%), and only 3% magnesium, whereas virtually all the fusible alkali is still present in the melt, and sodium grows proportionately from 2.7% to 4.5%. Silica likewise increases from 45.4% to 51.6% in this example.

Andesites, with even larger quantities of light-colored feldspar, is another common derivative of basaltic magma through fractionation.

If cooling proceeds and more crystals settle out of an ever-shrinking magma pocket, even more radical chemistries are reached. The last dregs of low-temperature magma, rich in alkali, will yield light-colored, siliceous lavas (over 65% SiO_2) known as *trachyte* and *dacite*, with as little as 3% iron and 0.75% magnesium oxides, whereas concentrated potassium and sodium range as high as 4 to 5%. Even more differentiated is *rhyolite*, with over 70% silica.

The changing composition of a volcano's lava through time therefore reflects the history of its magma: not only variations at the source, but also how the molten stock evolved during ascent, storage, and eruption remobilization.

Table 1.1 *Magma differentiation: a lava suite on Earth (Chaine des Puys)*

	Basalt (Cheire d'Aydat)	Leucobasalt (Puy de Dôme)	Trachyandesite (Puy de Pariou)	Trachyte (Puy de Dôme)
SiO_2	45.40	51.60	54.10	65.05
Al_2O_3	16.10	17.50	17.80	19.65
Fe_2O_3	7.12	4.69	7.70	3.25
FeO	4.49	4.78	1.10	0.01
MgO	6.90	4.20	2.65	0.75
CaO	10.00	7.40	5.70	1.25
Na_2O	2.70	4.50	5.00	5.05
K_2O	2.00	2.70	2.70	3.90
TiO_2	3.60	2.30	1.80	0.50
P_2O_5	0.64	—	0.70	0.10
MnO	0.19	0.21	0.20	0.15
H_2O^+	0.43	0.03	0.00	0.40
H_2O^-	0.10	—	0.25	0.10
Total	100.14	99.91	99.70	100.16

An example of magmatic differentiation on Earth: the lava suite of the Chaîne des Puys volcanoes in central France. The magma body underlying the volcanic chain provides a spectrum of flows, from basic and basaltic to acid and trachytic. The chemical trend is characterized by a concentration in silica, potassium and sodium (lavas lighten in color and become more viscous), while on the other hand the percentage of metallic oxides decreases (less iron, magnesium and calcium). Erupting temperatures are also lower across the lava suite.
From G. Camus, A. De Goer, G. Kieffer, J. Mergoil, and P.-M. Vincent, *Volcanologie de la Chaîne des Puys*, Aurillac, Clermont-Ferrand, 1982.

The eruptive process

Up until now, we have focused our discussion on the solid and liquid phases present in the magma, and the resulting diversity of the erupted lavas. But there is a third phase involved in magmas – the gaseous phase – which kicks in as soon as declining pressures allow volatiles in the batch to come out of solution.

Volatiles in a magma include water, carbon dioxide and monoxide, sulfur, argon, and halogen gases such as chlorine and fluorine. In their dissolved state, some of these volatiles already play an important role in boosting partial melting, since they lower the fusion temperature of silicates: if the Earth had suffered from a water-poor, 'dry' mantle, it would have experienced much less melting and volcanic activity.

FIG. 1.9. Lava fountaining at Pu'u O'o, Hawaii (June 1984). Fountaining is caused by the expansion of volatiles in the melt as it rises to the surface: above thresholds of 1 to 2% of water vapor or carbon dioxide, the basaltic magma is disturbed into a fiery froth that can rise to heights of several hundred meters. Note the lava flow in the foreground, spilling over from the crater lava lake. Credit: Photo by J. D. Griggs, USGS, Hawaiian Volcano Observatory.

As the magma oozes out of its matrix and starts to rise, volatiles come out of solution at specific pressure thresholds, nucleating as small bubbles. The bubbles then grow as the encasing pressure keeps dropping, all the way up to the surface.

As an example, let us follow the rise of a typical basaltic magma with one per cent dissolved water (by weight). Laboratory simulations show that small bubbles begin to exsolve out of the magma around one kilometer depth, and that they expand rapidly over the last hectometers. One hundred meters from the surface, the bubbles occupy three quarters of the available volume, and at that level of disturbance, the magmatic froth is blown to shreds. From ascent rates of one to two meters per second at one kilometer depth, the emulsion accelerates to blow-out speeds of one hundred meters per second and more in the last reaches of the erupting vent, lofting ash and magma droplets high into the air.

The eruptive behavior of a volcano is thus closely related to the magma's volatile content, as well as to the size and shape of the eruptive conduit. The example we just described – of a basaltic magma with one per cent water, pulverizing fine magma droplets high into the air – is refered to as an hawaiian eruption, since it is characteristic of Hawaiian volcanoes (see fig. 1.9).

Larger volatile contents (several per cent) drive their host magma into convective clouds of ash rising high above the erupting crater: these eruptions are named *plinian*, in reference to the Vesuvius blow out of AD79 described by Pliny the younger

FIG. 1.10. Plinian eruption at Mount Etna. When the volatile content of a magma greatly exceeds the disturbance threshold, the pulverized scoria and ash are lofted to great heights by the escaping gases. Air is sucked into the rising column, expanding in turn and driving the upward motion. Credit: Photo by D. Decobecq, LAVE.

(see fig. 1.10). When clouds of particulate matter are particularly dense, the eruptive column can also collapse under its own weight soon after leaving the crater, and spread as a base surge over the flanks of the volcano. These eruptions are named *pelean*, in reference to the deadly ash flow of Mount Pelée in Martinique (1902), and they embrace a variety of pyroclastic activities ranging from small *nuées ardentes* (Merapi type) to lateral *blasts* (Mount Saint Helens), and large *ignimbrite* ash flows (Mount Katmai).

At the opposite end of the scale, we find cases of low volatile concentrations in slow rising magmas, constrained by narrow chimneys. In such restricted magma columns, the few bubbles present rise faster than the magma itself, and sweep upward through the column, coalescing to form discrete 'gas packets' which break periodically at the surface. These rythmic belches of lava bombs, separated by periods of quiescence while the next gas bubble sweeps up the chimney, are known as *strombolian* eruptions, after the showcase Stromboli volcano.

Finally, if the magma is totally devoid of volatiles, it will flow peacefully out of the vent upon reaching the surface, without any noticeable disruption.

This range of eruptive behaviors, due to variable volatile contents, is one of the major characteristics of terrestrial volcanism.

With these notions in mind, we can now set about exploring in detail the volcanic features of the Earth, the Moon, and neighboring planets of our Solar System.

2 Volcanoes of Earth

A view from space

The exploration of planet Earth is still very much in progress, and regions long ignored are only today receiving their fair share of attention. We saw in Chapter 1 how geodesic surveys and submarine missions are bringing the deep ocean basins and underwater rifts into sharper focus. On land as well, the last frontiers are opening up to new mapping campaigns, with the assistance of satellite remote-sensing.

Satellite imagery helps us view the Earth in a more global fashion, and picture entire geological provinces and their tectonic relationships. It is perhaps no coincidence that the global model of plate tectonics was developed at a time of intense planetary self-scrutiny, in the decades of manned space flight and early space photography of planet Earth.

On a smaller scale, satellite imagery has helped to comb particular areas of the Earth in greater detail, and this has proved quite effective in locating uncharted volcanoes. An entire class of discrete, low-relief volcanic features – known as resurgent calderas – has blossomed since the advent of satellite imagery: not a year goes by without some new caldera being discovered, notably in Central and South America.

Satellites are also used to monitor active volcanoes, by relaying data from automatic stations installed on their slopes, and by imaging ongoing eruptions. Infrared imagery, in particular, is well suited to reveal heat anomalies over calderas and lava flows; and images taken periodically are used to map the growth of lava fields, and track ash plumes in the atmosphere.

The space era has also changed our perception of the Earth in an indirect way. Before the first space flights, the Earth was the unique model for geologists: its features had to be taken at face value, with no access to the surface of other planets for comparative purposes. Spacecraft exploration of the Moon, Mars, and Venus enlarged our vision by bringing out which features and processes were generalized throughout the Solar System, and which were unique to each planet.

In this respect, the Earth was found to stand out from its neighbors in one fundamental way: it is the only planet in the Solar System to reprocess its rigid surface through the slow ballet of plate tectonics, certain ocean basins expanding while others close up, their lithosphere curving back into the depths of the mantle. Volcanoes cluster both along the boundaries of these tectonic plates, and atop large-scale swells within the plates.

The Earth owes this unique, tecto-volcanic framework to two fundamental characteristics, which set it apart from other rocky planets. Firstly, the Earth is large – larger than Venus and much larger than Mars, Mercury, and the Moon – and we have seen in Chapter 1 how this brings about higher temperatures and prolonged conditions of melting at depth, a necessary condition for widespread and long-lasting volcanism.

The second major characteristic that shapes volcanic activity on Earth is the abundance of water. Water is present in liquid, gaseous, and solid form at the surface of the Earth, but also deep in the crust and mantle, where small amounts are dissolved in the mineral fabric – most notably inside garnets and amphiboles. By depressing the melting temperature of its host minerals, water enhances melting. For example, 'dry' mantle rock at 100 km depth would need to reach temperatures in excess of 1500 °C before starting to melt, whereas with 1% water, melting begins at 1300 °C. This 'wetness' of the Earth's mantle contributes greatly to the extent of magmatism at depth, and volcanism at the surface.

Underwater volcanism

Liquid water also covers two thirds of the Earth's surface, and the majority of eruptions occur on the deep ocean floor under 3000 to 4000 m of water, which corresponds to confining pressures of 300 to 400 atm. Abyssal eruptions are therefore very special, buffered by the weight of overlying water: gas in the erupting magma is generally prevented from expanding explosively by the high pressure environment, letting lava ooze out peacefully onto the ocean floor (one exception is when volatile contents are extremely high, as with some alkaline lavas enriched in CO_2, that manage to explode even at depths of 4000 m).

As an underwater volcano builds up to sea level, the eruption regime usually changes as the vents near the surface: the declining pressure allows trapped gas to expand and escape, fragmenting the magma to shreds (the resulting glasses are named hyaloclastites). Some of the most spectacular eruptions on Earth occur in this fashion, when a volcano peaks its nose above water, and both its magmatic volatiles and infiltrated seawater disturb the magma into phreato-magmatic explosions (Faial, Azores, in 1957; Surtsey, Iceland, in 1963).

Another trait of underwater volcanism, pointed out in the previous chapter, is how difficult it is to study and monitor. It takes sophisticated sonar surveys and submersible expeditions to get a close look at underwater features. But since they make up the largest family of volcanoes on Earth, we begin our survey of terrestrial volcanism with a visit to the mid-ocean ridges.

Mid-ocean ridges

Arguably, mid-ocean ridges are the most prominent geological features on Earth, although hidden from view by the blanket of ocean water. Early magnetic, gravimetric, and sonar campaigns identified the ridges as swells of volcanic rock, rising an average of 2000 m above the abyssal plains: they snake across the sea floor over a total distance of 60 000 km, more than circling the globe.

The underwater chain crosses the Arctic Ocean between Spitsbergen and Greenland, peaks above sea level to form Iceland, and dives back to an average

FIG. 2.1. Schematic
diagram of a section of
the inner rift valley and
adjacent walls on the
Mid-Atlantic Ridge,
showing the underlying
magma chamber and
dykes. The vertical
exaggeration is two-fold
(physiography by Tau
Rho Alpha). Credit: From
R. Hekinian *et al.*, 1976,
Volcanic rocks and
processes of the
Mid-Atlantic Ridge rift
valley, *Contrib. Mineral.
Petrol*, **58**, 83–110.

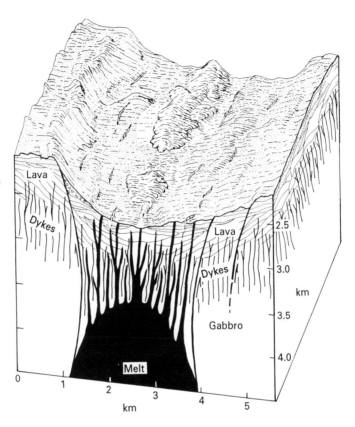

2000 m depth as it snakes down the middle of the Atlantic basin. Veering east around
Africa, the ridge continues across the floor of the Indian Ocean to a *triple junction*, a
fork where it splits into two branches: one section heads north to the Gulf of Aden
and up the Red Sea; while the other heads southeast around Australia and across the
southern Pacific, before resuming a northward course as the prominent East Pacific
rise, up to Central America where it ends in a series of faults.

On a regional scale the ridge is composed of segments, each a few hundred
kilometers in length, laterally offset by *transform* faults. This offset geometry reflects
the rupture pattern of the lithosphere when ocean basins rifted open, as well as the
dimensions of the rising diapirs which underpin each segment.

Mid-ocean ridges have a variety of profiles. Some ridges are broad with gentle
slopes and a rounded top, as in the East Pacific (fast-spreading ridges). Others are
narrower and steeper, and are grooved at the summit by an axial trough running the
length of the crest (slow-spreading ridges): the Mid-Atlantic Ridge, for example, has
an axial trough stepping down over a thousand meters to a central valley floor, four to
five kilometers wide (see fig. 2.1). In slow-spreading ridges, this faulted-down, axial
valley harbors essentially all of the ridge's volcanic activity. Magma permeates the
fractures of the arched-up crust, and collects in linear, shallow magma chambers,
that extend five to three kilometers below the surface, directly below the rift valley.
Most magma crystallizes at this level, along the walls of the cooling chambers, and

FIG. 2.2. Pillow lavas on the sea bottom near the Mid-Atlantic Ridge (Vemanaute exploration campaign). Lavas erupting on the sea floor develop a chilled skin that stretches and ruptures as the flow advances, giving the formation a typical bulbous appearance. Credit: © IFREMER.

slowly turns to gabbro rock (the coarse-grained, plutonic equivalent of basalt). Magma also permeates the fissures of the overlying crust, creating swarms of *feeder dykes* to the surface. Where the fissures intersect the ocean floor, the magma pours out as lava flows, and occasionally builds shields and mounds.

Lava behavior on the ocean floor is typically subdued, as we noted earlier, because the high pressures inhibit the expansion of volatiles. The ocean also acts as a chilling agent on underwater flows: erupting lava develops an envelope of quenched glass as it spreads onto the ocean floor, stretching and breaking its bulbous front to grow new lobes and 'toes'. This gives underwater lavas their characteristic rounded form, and their familiar name of *pillow lavas* – because their protuberances are the size and shape of large cushions (see fig. 2.2). Where the slope is steep, toes of lava break off into individual 'pillows', rolling down the abyssal slopes to collect as talus deposits.

Samples of deep-ocean lavas can be retrieved from surface research vessels by means of dredge nets hauled across the ridge or the abyssal plains. Other vessels – like the *Glomar Challenger* – lower and operate sophisticated coring equipment on the ocean floor, drilling hundreds of meters through sediment and crust to recover complete cores. Last but not least, spectacular diving campaigns have given geologists the opportunity to pick their own samples at depths in excess of 3000 m, with the assistance of floodlights and remote-controlled grabbing tools.

Samples recovered from mid-ocean ridges turn out to be *tholeitic basalts*, that are fairly enriched in silica, and are characterized by crystals of calcic to sodic feldspars

(40 to 50%); calcic and magnesium-rich pyroxenes (40 to 50%); and a sprinkle of olivine (around 5%).

Mid-ocean ridge basalts owe their distinct composition to the mode of magma generation at depth, a massive fusion of up to 30% of the rising mantle material. This high melting ratio explains why so much silica, iron, and magnesium is released into the melt. Inversely, early-fusing metals, like the alkalis, end up with low proportions (potassium oxide is down to 0.1%), diluted by the vast quantities of more refractory elements. In addition, the source mantle below ridges has a low alkalis content to start with, since alkalis were often removed during previous episodes of melting.

Spreading rates

The emplacement of fresh magma at mid-ocean ridges occurs episodically, with eruptions driven by the rate of replenishment of the magma chambers from the mantle below, and by processes within the chambers – mixing of different magma batches, exsolution of volatiles, and other physical and chemical disturbances.

During an eruption, magma rises through available fissures to spill onto the ocean floor; and when the eruption comes to an end, the last dregs of magma solidify in the feeder dykes. Further crustal extension is then needed for new fractures to accommodate the next batch of rising magma. In effect, the volcanic expansion of the ocean sea floor can be pictured as the growth of a deck of cards, standing on edge: extensional forces exerted on the deck make room in the center for the insertion of new 'cards' – sheets of magma – that turn to stone. Since the stack grows from the center outwards, the youngest dykes and lavas are found at the center of a mid-ocean ridge, and get progressively older away from it.

By measuring the age progression of lavas over large distances, one gets an estimate of the average eruption pace at the ridge – the *spreading rate*. Some ridges spread relatively slowly, as does the North Atlantic ridge, where the crust widens at a rate of 2 cm/year (the rate at which Europe separates from North America). Spreading rates are higher in the South Indian Ocean (6 cm/year), and highest in the Pacific, where part of the East Pacific rise spreads at a record 18 cm/year.

One must remember that these figures are only averages, derived over a sampling spread of several million years. A 2 cm/year rate does not mean that one sheet of magma 2 cm-thick pops up every year all along the length of the ridge. Instead, sizeable eruptions occur at irregular intervals – perhaps every century or thousand years on any one segment of the ridge – separated by periods of rest. Individually, each eruption is thought to create one or several dykes up to a meter thick, thus contributing meters of crustal extension at a time. It is only the average that boils down to centimeters per year.

Taking the average spreading rates of oceans, and integrating them along the entire length of the mid-ocean ridges (60 000 km), one can estimate that the average volume of magma emplaced each year reaches 15 to 20 km^3, which makes mid-ocean ridges by far the most active form of volcanism on Earth today. But then again, these are average figures: totals fluctuate randomly on a yearly basis; and on a much larger scale, the Earth also experiences long-term changes in sea-floor spreading, marked by reorganisation of plate motions, and 'jumps' in spreading ridge axes.

Exploring the rift

Diving expeditions using research submarines have given geologists unique opportunities to look close up at the morphology of mid-ocean ridges, and at the modes of eruption on the ocean floor.

In the course of the Franco-American FAMOUS campaign in 1974, research vessels dove 2700 m to the valley floor of the Mid-Atlantic ridge, southwest of the Azores Islands (37° N, 33° W).

The targeted segment displays a 'classic' Atlantic ridge profile, with a central rift 30 km-wide, stepping down 1000 m to an axial valley 3 km-wide. The submarines dove to the floor of the axial valley, where they photographed volcanic mounds and flows, and sampled chunks of basalt. The samples collected closest to the valley axis turned out to be the youngest (3000 years old), while those on the valley edge, 1 km to the side, yielded ages of 50 000 years, confirming the age progression away from the central axis of activity.

It is along this central axis that the divers observed the most conspicuous volcanic features: on the 1974 expedition, they circled a mound 1 km wide and 4 km long, rising 250 m above the valley floor. At each end of the mound, a stretch of relatively flat valley floor ran a few hundred meters to the base of the next mound.

Valley-floor volcanoes lack distinctive summit craters – at best they exhibit flat or slightly dimpled tops – and lack open fissures, clogged by underwater slumps and their own extrusive products. Flows of lava are prominent on their steep slopes, running dozens of meters in length, and fanning out in talus of broken-off pillows and fragmented volcanic glass.

Subsequent expeditions enlarged our vision of mid-ocean ridge volcanoes. Sonar-mapping campaigns delineated impressive arrays of cones and domes down the floors of the rift valleys: a 1988 expedition between 24° N and 30° N along the Mid-Atlantic ridge revealed 500 volcanic mounds over 50 m in height, strung along the 800 km length of the survey. Extrapolated over the total 60 000 km of mid-ocean ridges, this local count suggests that the population of central rift volcanoes could well reach 40 000 to 50 000 cones and domes.

Away from the central rift, older volcanoes occur on the ridge flanks and farther out in the abyssal plains. Most are subdued, due to tectonic deformation and slumping over the years, but one can still identify close to a dozen features per thousand square kilometers. Integrated over the total area of the ocean floors, the number of older volcanoes thus reaches the millions.

A difficulty arises when one attempts to pinpoint which underwater volcanoes are erupting today. It was almost a disappointment when the first submersibles descended to the Mid-Atlantic rift in 1974, only to find 3000 year old basalts. Some segments of the rift can indeed be dormant for millennia before undergoing a new pulse of activity.

FIG. 2.3. An active hydrothermal vent (left) at a depth of 3000 m on the East Pacific Rise (1982 Cyatherm exploration campaign). Nicknamed a 'black smoker', this hot spring spews forth 300 °C fluids rich in dissolved minerals. Hydrothermal circulation evacuates the heat of the newly formed mid-ocean crust, and plays a major economic role in concentrating precious ore deposits. Credit: © IFREMER.

As diving campaigns progressed in the late 1970s, remote-controlled probes with cameras were towed above the rift floor, in search of the most promising sites. In 1979, the Alvin crew dove on the East Pacific rise to discover a field of hydrothermal chimneys, spouting jets of sulfide and oxide solutions at temperatures over 300 °C. Since this landmark discovery, hydrothermal vents and chimneys have been photographed and sampled on a number of mid-ocean ridges, and testify to the intense seawater circulation through the cracks of the oceanic crust, convecting and carrying heat away from the top and sides of the magma chambers (see fig. 2.3).

An actual lava eruption remained to be observed on the deep ocean floor, but the scientific community got a significant boost in 1991, when Navy Intelligence allowed the civilian research agency NOAA to listen in to its vast network of espionage underwater microphones, spread over the ocean floor to track Russian submarines. In the course of their first year of surveillance, scientists at NOAA detected over 7500 seismic events in the Pacific Northwest, and on June 26, 1993, a swarm of earthquakes on the Juan de Fuca ridge alerted NOAA that a major eruption was underway. Rushing to the scene, the NOAA ship Discoverer lowered a two-ton undersea robot 2000 m down to the valley floor, where it was able to film an erupting gash 6 km long, spewing hot water plumes at one end and molten lava at the other. This was the first glimpse ever of the most active, and best-hidden form of volcanism on Earth.

Rifts above water

Volcanic rifting of oceanic crust does not all occur under water. There are a few rare spots on Earth where valley-floor volcanism can be observed above sea level, namely in Iceland where high eruption rates have raised the rift above sea level; and in the Afar region of East Africa, where an offshoot of the Indian Ocean Ridge tears into the continent.

A trek through the Afar basin is a magical tour through an ocean rift valley with its waters parted. The region is complex, due to its 'triple junction' tectonic setting. But the landscape of linear fissures and flattened cones is still a close analog to what occurs under 2000 m of water, at mid-ocean ridges. In the Afar, scientists were treated to a major eruption in 1978, when the Assal rift disgorged tholeiitic basalt, and the fissural extension of the crust reached 2.4 m in width.

Iceland also offers valuable insight into the mechanism of mid-ocean ridges: here, the North Atlantic ridge arches up above sea level, in the form of an island 100 000 km^2 in area, cut up into strips of progressively older lavas away from the active, central axis. The lavas farthest away, on Iceland's western and eastern coasts, are up to 20 million years old. The youngest lavas are erupted in Iceland's central rift, a broad swath of fissures 100 km wide, running northeast–southwest down the middle of the island. It is along these central grabens that one encounters most of the active volcanoes of Iceland, ranging from small fissures and shields to stratocones over 1000 m in height. The tallest volcanoes are covered by massive glacier caps, and their summits are flattened by the dynamics of subice eruptions.

On average, Iceland experiences one major eruption every five or six years, which makes it the second most active volcanic province on Earth today, behind Hawaii's Kilauea. With an area the size of New Hampshire, Iceland has contributed close to a third of all lavas erupted worldwide above sea level since the seventeenth century. Moreover, the island has hosted the largest lava eruption in human history: the Laki eruption of 1783. Activity lasted seven months – spread over 30 km of rift – with a discharge rate of 5000 cubic meters of lava per second, a flow rate comparable to that of the River Rhine. During this record eruption, which drove Iceland to the brink of famine and affected the climate worldwide, 12 cubic kilometers of fresh basalt were added to the surrounding countryside.

Iceland is exceptional in that it experiences eruption rates vastly greater than the already high averages of mid-ocean ridges. The Iceland segment has ballooned over the years into a voluminous volcanic plateau, 3000 m higher than the base level of the ridge.

Hot spots

Areas where large diapirs of hot, buoyant mantle rise to the surface are known as *hot spots*.

Hot spot provinces are conspicuous at the Earth's surface, because of the topographic arching and thinning of the crust above the rising plumes, the higher than average heat flow, and the great volume and distinct chemistry of their lavas.

Depending on the criteria used to define them, anywhere from 50 to 100 hot spots are presently transferring hot material from the lower mantle up to the surface, each rising plume measuring a few tens of kilometers in width. Plumes that have just reached the surface begin by erupting a partial melt rich in fusible elements (alkali basalt); while plumes that have been active for millions of years usually switch to voluminous outpours of less differentiated mantle melt (tholeiitic basalt). The distribution of hot spots around the globe shows several significant trends. Although a few plumes occur at, or near mid-ocean ridges (as in Iceland, the Azores, and the Juan de Fuca Ridge), most hot spots are independent of plate boundaries, and have a life of their own. Oceanic plates are pierced by hot spots in the Atlantic Ocean (Cape Verde, Canary, and Bermuda); Pacific Ocean (Hawaii, Tahiti, and Louisville); and Indian Ocean (Reunion, Marion, and Kerguelen Islands).

Continental plates are even more affected by hot spots. Although continental crust covers only a third of the Earth's surface, it overlies twice as many hot spots than does oceanic crust. This concentration is most spectacular in the case of Africa, which alone comprises over a dozen arched-up swells with anomalous heat flow: a first swath of hot spots runs down the continents eastern margin from Ethiopia to Mozambique, fueling the East African Rift; and a second cluster of hot spots runs from Libya and Tchad in the north to Nigeria and the Gulf of Guinea in the south, including the Hoggar plateau – a crustal swell 1000 km in diameter – and the 4000 m Mount Cameroon.

Hot spot diapirs are also located beneath Arabia (close to the Red Sea Rift); under the Madagascar crustal block; China; Lake Baïkal and Siberia; and in North America under the coast of Alaska, the Yellowstone volcanic plateau, and the Colorado plateau. On the other hand, South America and Australia only boast a couple of marginal hot spots, and Europe fares little better with only two suspected plumes, one under Germany's Eifel volcanoes and the other beneath France's Massif Central.

The great African rift

The concentration of hot spots under continental crust deserves a closer look, for it suggests some form of causality.

Continents are thick caps of silica-rich material, assembled over time as large-scale fusions of the upper mantle concentrate siliceous and aluminous material at the surface. Because of its low density, this crustal material is prevented from sinking back into the mantle, and aggregates instead to form continental 'rafts'. Harder to rupture than oceanic lithosphere, this thick continental crust acts as an insulating lid over the mantle, driving temperatures up and triggering instabilities at depth: this might explain why so many hot spots are found below continents.

At the surface, the evolution of a continental hot spot follows a characteristic pattern. The expansion of the hotter mantle uplifts and stretches the lithosphere, to the point where normal faulting gets underway, and the thinned-out crust collapses centrally, creating tilted blocks and grabens. Magma ascends and seeps through the faulted crust, flowing out onto the graben floors.

This scenario is presently being played out under the East African Rift, a faulted-down corridor roughly 100 km wide that stretches 4000 km through Ethiopia, Kenya, Tanzania, and into Mozambique, peppered with shield volcanoes on its floor and along its rims: Longonot in Kenya; dormant Kilimanjaro towering 5930 m over Africa (see fig. 1.7); Nyamlagira and Nyiragongo; Ngorongoro; and Oldonyo Longai.

The thermal arching-up, extension, and central collapse of the East African province began 35 million years ago, with a sequence of massive eruptions: in Ethiopia, extensive fields of lava date back to this period, and stack up in layers totalling over a kilometer in thickness.

As is the case with most incipient hot spots, these early East African lavas are alkaline basalts, enriched in sodium and potassium. Such alkaline magmas, if they get a chance to decant in shallow magma chambers, might also yield phonolites (poor in silica) and trachytes (rich in silica). Where crustal material is melted down, silica-rich rhyolites can also be erupted. Trachytes are more viscous than basalts but still flow to some extent; whereas phonolites and rhyolites are so rich in gas and silica that they usually burst out of their magma chambers in clouds of incandescent material to roll over the landscape and settle as pyroclastic deposits – ash flows and larger ignimbrite fields.

Besides the classic calc-alkaline suite, the onset of hot spot activity under a continent can also yield uncommon lavas, as observed on the margins of the East African Rift. Volcano Oldonyo Longai in Tanzania is famous for erupting a chocolate-coloured melt of calcium and sodium carbonate: very fluid at its erupting temperature of 500 °C, this original magma solidifies to a whitish rock, not unlike limestone. Named carbonatite, it probably owes its unique composition to the interaction of the rising magma with the limestone beds of the uppermost crust (see fig. 2.4).

Alkaline basalts and their derivatives mark the onset of hot spot activity under a continent. But as the arched-up crust stretches and thins out, more mantle material is brought closer to the surface, and with the drop in pressure melts much more massively than the earlier, deeper batches of melt. Alkaline lavas are followed by these large-fusion, tholeiite basalts, which are close in composition to those erupted on the ocean floor.

When an arched-up rift thins out over a hot spot, its fragile crest ultimately founders, and with time the faulted-down rift drops below sea level and floods with water, to become an embryonic ocean.

The East African Rift is an interesting example of how hot spot activity leads to continental break-up and the genesis of a new ocean. It occupies an interesting position in the plate tectonic map of the world, being one of three branches of a star (a tectonic triple junction). The two other branches have already collapsed below sea level and are spreading in ocean ridge fashion: they are the Red Sea (northern branch), and the Gulf of Aden (eastern branch).

FIG. 2.4. Oldonyo-Longai ('the mountain of God'), west of Kilimanjaro in Tanzania. The caldera (top) is floored by fluid flows of white carbonatite, a rare form of lava composed of sodium and calcium carbonate. These low-temperature (500 °C) fluid lavas are encountered in continental rift zones and might originate from the melting of limestone basement rock. Note the central spatter cones, surrounded by the most recent, darker flows. Carbonatite flows (bottom) are indeed black upon eruption, before turning white soon after solidifying. Credit: Photos by Alain Catté (top) and B. Demarne (bottom), LAVE.

The East African branch is itself reaching maturity, the crust thinning out over the hot plume, and the mantle melting pervasively, close to the surface. As a result, one finds in the East African Rift both alkaline lavas from the deep hot spot source, and patches of tholeiitic basalt, from the spreading upper mantle melt.

Presently, the foundering floor of the East African Rift is sprinkled with alkaline and acidic lakes, but if the stretching and foundering persist at the present rate (a couple of millimeters a year), the rift will ultimately follow the fate of the two other branches, flooding with seawater and slowly splitting the eastern side of East Africa away from the rest of the continent. This change of geography will not happen overnight: at one millimeter of stretching per year, it will take 10 million years for the East African Rift to open up by another 100 km into an embryonic ocean.

Flood basalts

Not all continental rifts become spreading oceans. But the tectonics of stretching over hot spots does guarantee massive melting and the easy access of magma to the surface. The greatest lava fields on Earth are often associated with hot spots that were active in past episodes of our planet's history.

These giant lava fields are called *traps* – the Swedish word for stairs, as they are characterized by terraced lava layers – or *flood basalts*. The most famous one is the lava pile that makes up the Deccan plateau around Bombay and stretches down the western coast of India. The Deccan Traps cover at least 500 000 km² (the area of a country like France), and in places the lava layers reach 2 km in thickness. The total volume of lava probably exceeds one million cubic kilometers, erupted over a time period of some 500 000 years (65 million years ago).

Another famous trap is the Parana lava field in South America, crossing from Brazil into Uruguay and Paraguay. Here as well, estimates call for an area close to one million square kilometers, and a total volume exceeding one million cubic kilometers. The lavas were erupted in a time interval of a few million years at most, when hot spot rifting was at work separating South America from Africa (130 million years ago).

There are 12 episodes of flood basalts or traps, that took place over the last 250 million years, including the traps of Siberia (250 million years old); the eastern seaboard traps of North America and the Karoo basalts of South Africa, which marked the continental rifting that opened up the Atlantic Ocean (200 my); the already mentioned Parana (130 my) and Deccan (65 my) Traps; the younger Brito-Arctic Traps (62 my); the Ethiopia lava fields at the onset of the East African rifting (35 my); and most recently the lava fields of the Columbia River Plateau (17 my) in the American Northwest, the last trap eruption on record.

To get a sense of the magnitude of trap volcanism, one should note that an average trap represents one million cubic kilometers of lava erupted over one million years – an average rate of one cubic kilometer per year if spread out evenly over the time interval. But trap eruptions are certainly much larger than this average figure, separated by centuries of quiet. Indeed, some individual trap flows reach 50 to 100 km in length and several thousand cubic kilometers in volume, making them truly gigantic events. They would dwarf the largest lava flows witnessed by man – such as the record 12 cubic kilometer flow erupted by Laki (Iceland) in 1783. Indeed, whereas Laki erupted at a rate of one ton of magma per second and per meter of fissure, trap rates are believed to be twenty to thirty times larger.

On account of their exceptional discharge rates, trap eruptions do not build up mounds, shields, or other volcanic constructs of that sort: trap lavas stay so hot and fluid that they follow gentle slopes and spread thinly over large areas, often covering up the very fissures that discharged them in the first place. Traps are characterized by this absence of recognizable eruptive vents, a feature they share with similar lava fields on the Moon, Mars, and Venus.

Despite their cover-up, the inner workings of trap eruptions are revealed in exposed outcrops at several places on Earth. At the Columbia River Plateau, eruptive fissures now stand out as long dykes of chilled magma intersecting the surface, hundreds of meters long and up to a meter in width. In one such area of the plateau, nearly twenty thousand dykes have been mapped: this serves to illustrate the extent of crustal extension that accompanies trap eruptions on Earth.

Plains volcanism

When eruptions take place on Earth at more modest rates, there is a change in the type of landscape created at the surface. Lower discharge rates through narrower fissures will impose severe cooling and flow constraints, and only a few individual feeding pipes will remain active rather than entire fissures. The discharged lavas, less voluminous and quick cooling, build shield-shaped aprons around the central vents – shields which often overlap, and funnel their lava flows in the 'saddles' between them.

This type of volcanic behavior and morphology has been defined as *plains volcanism*, and its prime example is the Snake River Plain of Idaho (see fig. 2.5). There, on the margins of the Columbia River Plateau, dwindling magmatism shaped a landscape of overlapping shields, averaging 5 to 10 km in diameter, with low-incline slopes (0.5°), and slightly steeper summits, dimpled with pit craters (see fig. 2.6).

In such provinces, individual lava flows fed from pit craters and open fissures are commonly 5 to 10 m thick (whereas the flow 'sheets' of trap eruptions are up to three times thicker). Some are narrow enough to build overcasing roofs of solidified basalt: named *lava tubes*, these insulating pipes can funnel lava flows dozens of kilometers downslope, before cooling brings them to a halt.

Plains volcanism therefore promotes a pile-up of overlapping shields interlaced with lava tubes and flows, in contrast to the large continuous sheets of trap eruptions. These morphological differences are due fundamentally to contrasting discharge rates – high versus low magma output.

FIG. 2.5. Schematic block diagram of 'plains' type volcanism, based on the example of the Snake River Plain of Idaho. Low eruptive rates in a continental tensional environment lead to a pile-up of fissure-fed flows and discrete low shields, overlapping and burying each other. Credit: From R. Greeley, 1982, The Snake River Plain, Idaho: representative of a new category of volcanism, *J. Geophys. Res.*, **87**, 2705–12.

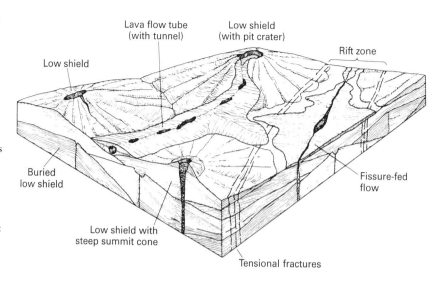

Oceanic hot spots

The formation of traps and lava plains is conditioned by complex crustal uplift and thinning of the continental crust. Oceanic hot spots have an easier time breaking and infiltrating the lithosphere with magma: firstly because oceanic plates are thinner than continental plates (10 to 50 km on average, versus 100 to 200 km for continents); and secondly because oceanic plates are laced with 'pre-fabricated' fractures, which they owe to their growth in a ridge environment truncated in segments. These fracture patterns lacing the oceanic plates may be utilized by rising magma batches.

When a plume rises under an oceanic plate, it therefore meets little resistance in arching up the thin crust and investing its faults. The first flows to spill onto the ocean floor are usually short – probably because the initial eruptions are of short duration – and underwater shields start out fairly steep as a result. As they grow closer to the surface and peak above water, their eruptive regime switches over to more explosive behavior on account of the declining pressure, with large amounts of fragmented volcanic glass generated by steam explosions and lava flows entering the sea. These unstable talus deposits keep the slopes fairly steep. After their tumultuous emergence above sea level, shield volcanoes develop gentler slopes, as their subaerial, little-disturbed lava flows grow longer and spread out over wider areas: from 30 degrees underwater, slopes commonly flatten out to 5 to 10 degrees above the surface.

Réunion Island, in the Indian Ocean, is a good example of an oceanic shield built over a hot spot. The construct has a 240 km-wide base (as measured on the ocean floor), rises 4500 m to sea level, and an extra 3000 m above it. The Piton de la Fournaise eruptive complex at the summit of the island is classified as one of the most active volcanoes on Earth today.

FIG. 2.6. A typical landscape of plains volcanism. Sand Butte, Idaho, is a cinder cone 1200 m in diameter straddling an eruptive fissure. Cones of the sort are characteristic of fissure-fed eruptions switching to localized pipes that end up focusing the activity. Notice the texture difference between the smooth ring of ejecta and the rougher, rope-like lava flows surrounding it. Credit: From R. Greeley, 1982, The Snake River Plain, Idaho: representative of a new category of volcanism, *J. Geophys. Res.*, **87**, 2705–12.

The Hawaiian Islands are another prime example of oceanic shields built over an active hot spot. The main island reaches 10 000 m in height above the sea floor (5000 m to sea level, and 5000 m above it), and estimates of the shields' volume range anywhere from 40 000 to 400 000 km^3, depending on which crustal level is defined as the base.

The chemistry of oceanic, hot spot magma is very similar to that of continental, trap magma: in both cases, hot spot magma originates in the deep mantle (between 100 and 400 km depth), and starts off rich in fusible elements. The first lavas to emerge in an oceanic hot spot environment will be alkali basalts – as in incipient continental hot spots – with relatively high proportions of potassium, sodium, and also a good deal of titanium.

A change in chemistry occurs when the hot spot plume begins to melt the surrounding upper mantle on its way to the surface: massive melting of this encasing mantle – between 100 and 60 km depth – then takes over the eruptive process, and the lavas come out looking like mid-ocean ridge, shallow-genesis basalts. *Olivine tholeiite* – which gives its color to the dark, emerald green beaches of Hawaii – falls into this category.

The Hawaiian hot spot

The shields of Hawaii's Big Island are presently in full growth, spewing forth flow after flow of tholeiitic lava. The Mauna Loa shield averages one eruption every five years, and the Kilauea complex on its southern flank is even more active with no less than fifty recorded eruptions so far this century. The latest eruption to date in Kilauea's eastern rift zone has proceeded virtually uninterrupted from 1983 to the year of writing (1995), with lava flows running down to the coast through lava tubes.

Lava tubes are a common fixture of shield volcanoes: it is estimated that close to 80% of all lava flows on Hawaii grew in this fashion, and one can picture the islands as massive plumbing systems of stacked, basaltic pipes.

The eruptive regime of Hawaiian shields is essentially peaceful: *hawaiian* eruptions involve fluid, basaltic magma, bearing 0.5% to 1% volatiles (mostly water). The fluid magma is slightly disturbed by gas expansion and venting as it reaches the surface, especially at the onset of eruptions, when there is enough gas pressure to propel magma droplets hundreds of meters high: these spectacular displays are named lava fountains, or curtains of fire when they spread along a fissure. When the feeding magma pocket is sufficiently degassed, lava fountains subside and the magma disgorges peacefully out of the vents.

Magma can also pool inside a vent or a caldera, giving birth to a lava lake, with or without exit channels to feed lava flows and tubes. Calderas which were once occupied by lava lakes mark the summit of Mauna Loa. Kilauea has its own caldera, a circular downdrop three kilometers wide and floored by solidified lava flows and pools (with a nested pit several hundred meters wide, named Halemaumau). A great, incandescent lava lake occupied the caldera floor from 1824 to 1923, before the magma drained back.

Island chains

Up until now, we have reasoned as if volcanism was a stationary event, in space and time. This is not the case on Earth, where volcano-bearing crust is almost everywhere in motion relative to the hot mantle beneath: the drift rate is 10 cm/year in the Central Pacific. Relative to the underpinning Hawaiian hot spot, the Pacific lithosphere is therefore carrying volcanic features on its back away from the fixed magma source at a rate of 10 cm/year – 1 km per 10 000 years. Since a plume of hot mantle is typically 100 km in width, volcanoes popping up in the center will drift out of its zone of influence in about a million years, and eventually 'shut down'.

On drifting plates it is therefore not one giant volcano that builds up over a hot spot, but a chain of medium-sized shields, lined up from the hot spot in the direction of plate drift. The active volcano of the chain is the one growing over the hot spot;

FIG. 2.7. The Hawaii-Emperor seamounts island chain was formed by a hot spot active for over 70 million years. Since the Pacific plate is in motion relative to the deep-anchored hot spot, volcanic edifices built over the magma source are carried away atop the plate, first toward the north, then – as plate motions readjusted 44 million years ago – toward the northwest. The line-up of seamounts, shoals and islands indicates the drift direction, with the inflection point near Kimmei seamount. The active island of Hawaii (lower right) is the latest volcanic island to build over the hot spot. Credit: From D. A. Clague and B. Dalrymple, 1987, courtesy of David Clague.

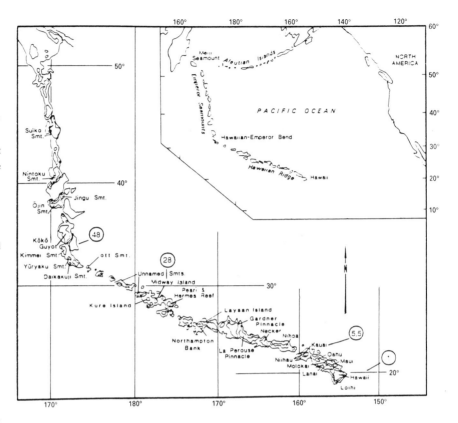

other volcanoes display progressively older ages the farther they are from their place of birth.

The Hawaiian Islands are a classic example of a hot spot island chain (see fig. 2.7), with the plume anchored at 20° N and 155° W. The Pacific plate drifts at an average rate of 10 cm/year over the hot spot, in a northwesterly direction. Directly above the hot spot, active volcanism is presently building the shields of Mauna Loa and Kilauea – the Big Island of Hawaii. In the 'wake' of the Big Island, older, eroded volcanic islands line up to the northwest: first is Maui and its young volcano Haleakala. The main shield of Haleakala was built over the hot spot one million years ago. Since then, the drifting Pacific plate has carried the island 100 km west to the edge of the hot spot, putting Haleakala on the brink of volcanic extinction.

Beyond Maui come a few smaller islands (including Lanai and Molokai), then Oahu and Kauai. The latter is 500 km from the hot spot, and its extinct, deeply eroded shields are 5 million years old. Beyond the Hawaiian Islands, the chain changes names but the alignment continues, with the French Frigate shoal, 1000 km from the hot spot, ranking 10 million years of age; and Midway Island, 2000 km distant and 28 million years old.

At the 3000 km mark, the shields are now submarine and make a bend in their alignment, striking due north across the progressively older sections of the Pacific plate, straight to the trench near the Kamchatka peninsula. This remarkable change in strike tells us that the motion of the Pacific plate started off due north relative to the hot spot (close to 80 million years ago), but that the plate changed direction about 44 million years ago: at the level of Kimmei seamount, the alignment switches to its present northwest/southeast trend, straight to Hawaii. Island chains that developed over 'fixed' hot spots are thus a precious marker of plate drift rates and directions over the past hundred million years.

Subduction and island arcs

At mid-ocean spreading ridges, the creation of new crust averages 1 to 2 km^2 per year (10 to 20 km^3 in volume, if one takes the average thickness of the crust to be 10 km). It follows that our planet should increase in size if no other mechanism came to balance crustal growth at the ridge: with an added 2 km^2 per year, the area of our planet would indeed double in a mere 250 million years. We know this not to be the case: the area of the Earth has remained constant over billions of years, which means that somewhere on the planet, mechanisms of destruction are doing away with old lithosphere at about the same rate that new lithosphere is created at the ridges.

These destruction zones can be spotted around the globe as deep ocean trenches at plate margins – most notably around the Pacific basin, in the Indian Ocean, and in the Caribbean. It is along these trenches that old and dense lithosphere curves back into the viscous mantle, where it is melted and 'recycled' – a process known as *subduction*.

The subduction process can be followed in great detail, thanks to the earthquakes triggered by the sinking plates. Seismic maps of subduction zones clearly show the contours of the descending slabs down to 300 km, and occasionally much deeper: the record of the deepest subduction earthquake belongs to the Celebes trench in the Western Pacific, with a 720 km-deep event in 1924. Where earthquakes end is where the subducted slab reaches such temperatures and pressures that it no longer behaves rigidly, and instead blends plastically into the hot mantle.

The process of subduction is reinforced by phase changes in the sinking slab. Oceanic plates are made up of basalts, gabbros, and sub-basement mantle rocks rich in olivine, pyroxene, amphibole, and feldspar. As these mineral assemblages sink into the mantle and meet rising pressures, their crystalline structures reorganize into more compact, denser arrangements. Feldspars turn into spinels at pressures of 300 million pascals (20 km depth), and spinels shrink in turn to garnets at pressures of one billion pascals (70 km depth). In so doing, they raise the density of their host rock – gabbro or basalt – from 3 to 3.5: the denser rock is called *eclogite*, and the slabs gain momentum, sinking ever more readily in the surrounding mantle.

Along with seismic energy, much frictional heat is released when a lithospheric plate is subducted into the mantle. Shearing and melting occur in the contact plane between subducting and overriding plates, and melting there is enhanced by the mobilization of hot fluids, flushed out by the heating and dehydration of the subducting slab. Magma produced in this complex environment finds its way to the surface through the cracks opened up by extensional tectonics over the bending, subducting slab. These magma diapirs end up intersecting the surface on a line parallel to, and behind the trench – a volcanic arc. The volcanic alignment is called an *island arc* when it occurs in an ocean basin, where two oceanic plates converge; and a volcanic *cordillera*, where an oceanic plate dives under a continent and builds volcanoes backshore, behind the coastal trench.

The Caribbean arc, with its historically active volcanoes of Mount Pelée in Martinique, La Soufrière in Guadeloupe, and La Soufrière of Saint Vincent is an example of an island arc; the long chain of volcanoes that make up the Andes, where the Pacific plate dives beneath South America, is an example of continental cordillera.

Subduction volcanism

Subduction volcanism is the most varied, and also the most explosive form of volcanism on Earth. Subduction volcanoes are dangerous: their outbursts are responsible for some of the worst natural catastrophes in human history. Over the past two centuries, Tambora volcano heads the infamous list with 92 000 victims in its explosive eruption of 1815; followed by Krakatoa in 1883 (36 000 casualties), both in Indonesia; Mount Pelée of Martinique in 1902 (29 000 casualties); and more recently the eruption of Nevado del Ruiz, Colombia, in 1985 (25 000 casualties).

Explosive volcanism in subduction environments is caused by the strong concentration of dissolved volatiles in the rising magma (water, carbon dioxide). When they reach the surface, these viscous, siliceous magmas are often blown apart by their fast-expanding gases, leading to explosive outbursts of plinian ash plumes and pelean ash flows.

The main lava family in subduction settings consists of the gray-coloured *andesites* – named after the Andes cordillera, where they abound. Andesites are also the main lava constituent of island arc volcanoes. This characteristic line of lavas ranges from *andesitic basalts*, at their most undifferentiated stage (52% SiO_2), to *andesites* (60% SiO_2), and more differentiated, siliceous, and volatile-rich end-members: *dacites* (65% SiO_2), and *rhyolites* (over 70% SiO_2).

Sediments might play a role in the magmatism of subduction zones. Trenches are usually clogged with sediments, scraped and stacked in the fault zone by the sliding, underriding plate: these are both sediments accumulated over the years on the oceanic plate on its way to the trench – mostly calcareous and siliceous tests of plankton and other forms of marine life – and detrital sediments shed off the nearby continents. When they reach a subduction trench, carried piggyback on the creeping plate, most of the sediments are sheared off by the 'grating edge' of the opposing, overriding plate and remain at the surface, piling up over the trench in fan fashion.

A fraction of the load, however, passes the grating filter and accompanies the crustal slab down into the hot mantle, especially if the plate develops graben as it flexes, trapping sediment in these structural 'drawers'. Because they contain abundant hydrated and carbonated minerals, sediments will exsolve volatiles when they heat up: added to the volatiles from altered lavas and serpentinites, these will act to depress the melting point of the slab and of the mantle around it. The wedge of mantle rock over the subduction plane is particularly sensitive to this injection of volatiles, and generates most of the magma that rises to the surface in subduction environments.

The rate of convergence and the angle of dip of a subducting plate will affect the nature and extent of magma production. Rates of convergence are as varied as spreading rates at the mid-ocean ridges – from a couple to a dozen centimeters a year on average. Gravity will make a slow-moving slab dip steeply (up to 80°) into the mantle over a short horizontal distance, whereas a fast-moving slab will slide at a shallower angle (down to 30°) under the over-riding plate, and cover a great horizontal distance before it is deep enough to melt.

Different subduction geometries will yield different regimes of melting. At the Marianas trench, for example, an oceanic plate dives at a fast clip under another oceanic plate: few sediments are involved, and the shallow mantle wedge between subducting plate and surface yields a basaltic magma (quartz tholeiite), with very little calc-alkaline products. Behind fast-subducting slabs, the magma might even come to resemble that of a mid-ocean ridge basalt (olivine tholeiite), especially where extensional tectonics rip open the crust over the bending slab, to create back-arc basins. This is notably the case behind the Japan trench (extensional basin of the Sea of Japan), and the Kuril Islands trench (back-trench Sea of Okhotsk).

Cordillera volcanism

Along the western coast of South America, the Nazca oceanic plate dips steeply under the American continental plate, setting off large amounts of calc-alkaline volcanism where the subducted slab – and the mantle wedge above it – undergo melting. Along the trench, the downgoing slab is not dipping everywhere at the same angle beneath the continent – the slab is cut up in multiple tongues as it were – and the influence of dipping angle is particularly evident from place to place: volcanism is strongest where the dip angle is greatest, resulting in deep-source, alkaline magmatism comparable to that of an incipient hot spot; whereas volcanism is weak to absent where the slab underthrusts the continent at a shallow angle.

Island arc and cordillera volcanoes are observed to line up at fairly regular intervals along the strike of a trench: the spacing is that of the diapirs rising from the melt zone. In the Aleutian arc, volcanoes are spaced every 50 km on average; in the Andes, back-trench volcanoes are lined up in segments 100 to 300 km long, offset by fault zones of high seismicity.

This overview would not be complete without a look at the special case of continental collision. When a subducting oceanic plate carries a low-density continent on its back and brings its rafting passenger to the trench, the buoyancy prevents the continent from going under. If the over-riding plate also carries a continent, the result is a continent–continent collision – a suture zone where the granitic and sedimentary material from both plates crunch up in accordian fashion, and the imbricated continental wedges thicken the local crust. Complex tectonics come into play, and extensional zones develop, that can harbor volcanism.

The Himalayan range was created in just this fashion: an oceanic plate was subducting 'normally' under Asia until 60 million years ago, when it brought India on its back up to the trench. The leading edge of the Indian continent was forced down into the trench, but the plates quickly locked horns, and subduction all but stopped in the suture zone. In the back country of the jammed trench, extensional tectonics accommodated some of the stress, and allowed magma to rise from melting 'scales' of continental crust at depth (Tibetan plateau and Baïkal rift zone). Similarly, volcanism developed behind the collision front of the Turkish and Asian plates, giving rise to the calc-alkaline volcanoes of Iran and Turkey in a back-trench, extensional environment.

Ignimbrites and calderas

Subduction volcanoes owe their explosive personality to abundant volatiles, dissolved in siliceous magma: at the top end of the scale, rhyolitic magma (over 70% SiO_2) can contain up to 7% water in solution. One then gets a hot emulsion of ash and gas: at modest eruption rates, ash columns will climb high in the atmosphere (*plinian* eruptions); and if the discharge rate is particularly high, the dense columns of ash will collapse under their own weight upon the flanks of the volcano, spreading downslope to form *nuées ardentes* (*pelean* eruptions).

Volatile-rich eruptions do not always revolve around ash columns blowing out vertically from central craters. In many cases, volcanic chimneys are obstructed and sealed, and volatile pressure mounts in the constrained magma. Domes and cones can then tear along preexisting fractures, and blow clouds of ash sideways – dangerous events known as *blasts*. Mount Saint Helens, in the American Northwest, demonstrated the mechanism of a blast on May 18, 1980, when the north flank of the stratocone collapsed in a gigantic landslide, and the sudden drop in pressure led to an explosive expansion of volatiles. A lateral blast shot several cubic kilometers of hot rock emulsion rolling across the countryside, while a column of ash rose vertically from the gutted volcano, and reached an altitude of 27 km (see fig. 2.8).

Another famous back-trench eruption is the nuée ardente of the Mount Pelée of Martinique: on May 8, 1902, after several days of volcanic unrest, a pyroclastic flow shot out of the faulted mountain face and rolled down the valley to engulf the town of Saint-Pierre, killing its 29 000 inhabitants. This dreadful ash flow was the first of a series: in 1902 and 1903, close to 60 pyroclastic flows ran down the mountainside, each dropping thousands of cubic meters of siliceous ash.

However spectacular, these historical nuées ardentes are minuscule in comparison to the major pyroclastic flows preserved in the geological record. These major flows are not thousands but millions of cubic meters in volume, and are called *ignimbrites*. Their shards of rock and glass are often fused together, giving them a lava flow appearance, although they were emplaced from turbulent, ground-hugging clouds of ash and gas.

Only one pyroclastic event of ignimbrite magnitude took place in recorded history, fortunately in an unpopulated area: this was the Katmai eruption of 1912, in Alaska. Spotted by an aviator several days after the eruption, the Katmai pyroclastic

FIG. 2.8. The eruption of Mount Saint Helens in the Cascades range (USA). The 1980 eruption began with a gigantic landslide of the volcano's northern flank, violently decompressing the entrapped volatiles: the lateral blast felled trees 25 km away from the crater. The eruption then switched to plinian activity, with a column of ash and gases rising 20 000 m. Credit: US Geological Survey, Cascades Volcano Observatory.

flow funneled down a valley to pile millions of cubic meters of hot tuff in an elongated arena 20 km long by 10 km wide. Five years after the event, a team of explorers visited the ignimbrite blanket, still steaming with vapors, and named it the 'Valley of the Ten Thousand Smokes'.

The Katmai ignimbrite eruption was not directly observed, but the explosive sequence was recorded by distant seismometers: from these records it appears that the eruption lasted three days. Later studies revealed that the pyroclastic flows blew out of a fracture zone at the base of Katmai volcano, and that the base surges were followed by the tearing open of the dome, and a powerful ash and pumice blast into the atmosphere – a combined volume release of 25 km³. Where the volcano once stood, the violent blow-out left a circular depression in its place – a *caldera*.

Large calderas are typically associated with ignimbrites. They mark the collapse of the ground surface over magma chambers emptied by catastrophic eruptions. In island arc settings, the triggering mechanism can be the infiltration of seawater in the volcano's plumbing system: both the Tambora (1815) and Krakatoa (1883) catastrophies – the two most violent eruptions of modern history – were due to magma/seawater interaction. In both cases, the volcanic islands were disintegrated by the explosions, Tambora holding the record with 100 km³ of tephra blown sky high. The explosion and foundering of Krakatoa, nearly as violent, led to the underwater collapse of the volcano, triggering a tidal wave that caused 36 000 casualties on the nearby shores of Java and Sumatra. Further back in time, the phreato-magmatic explosions of Thera-Santorini volcano in the Aegean Sea (1500 BC), and of Taal volcano in the Philippines likewise left impressive calderas.

Resurgent calderas

The largest calderas on Earth are caused by the collapse of their emptying magma chambers during an eruption, rather than by explosions in the firecracker sense of the word. The Katmai caldera resulted from such an implosive collapse; and so did all the large calderas of prehistoric time. Crater Lake in Oregon is a good example of caldera collapse. It was formed by the catastrophic eruption 7000 years ago of a large volcano, which foundered as the drained magma reservoir gave way beneath it. The result is a deep bowl 10 km wide, now occupied by a lake. The Taupo caldera in New Zealand is even larger, reaching 30 km in diameter: it was created by the vertical expulsion of vast quantities of ash, leading to the collapse of the magma chamber.

But the record belongs to Toba caldera, in Sumatra: the 100 km by 30 km formation results from the cataclysmic eruption, 75 000 years ago, of 2800 km^3 of rhyolitic magma. This extraordinary eruption is believed to have taken place over a two week period, starting with a vertical, plinian eruption of 800 km^3 of ash (one thousand times the output of the Mount Saint Helens blast), followed by waves of pyroclastic flows jetting out of fault rings in all directions, and closing with the giant collapse of the emptied magma chamber. Up to 2000 m deep, the caldera is now flooded by a lake. Since its paroxysmal eruption, Toba has evolved: over the millennia, magma has again crept up through the fissures and inflated the rocky substrate. A lava dome now rises at the center of the lake.

This cyclical behavior is characteristic of a special class of calderas – *resurgent calderas* – which stock magma for hundreds of thousands of years, grow unstable, and blow out all of their ash and gas in a matter of days or months, before beginning a new cycle of inflation. Resurgent calderas are found in subduction settings – island arcs and continental, back-trench provinces – and occasionally in hot spot environments, and continental rifts. Resurgent calderas are associated with large magma pockets of volatile-rich, crustal material (see fig. 2.9).

Despite their large size, resurgent calderas are difficult to locate. Their relief is subdued, and virtually all are inactive at present, although many are slowly building up magma at depth. Erosion of their soft ash beds adds to their cover-up. Resurgent calderas were poorly known until the advent of satellite remote-sensing, when giant tuff beds fell under the scrutiny of multispectral cameras. In false-color renditions, ignimbrite fields stand out in pastel colors (on the ground, ignimbrites are known to have gray, tan, or buff hues), and their soft strata are often deeply gouged by erosional channels, displaying dendritic or reticulate networks (see fig. 2.10).

Altimetry surveys show most resurgent calderas to be subtle, gently-sloping shields, with depressed central areas, and the tell-tale, resurgent 'heart' in the middle. Other features of resurgent calderas include alignments of small lava domes along arcuate ring faults – faults which are both the fissure zones of major eruptions, and the slip planes along which the caldera subsides.

FIG. 2.9. Cross-section of Long Valley caldera, California. This resurgent dome has been the theater of massive ignimbrite eruptions in the past, such as the 700 000 year old Bishop tuff. Seismic tomography presently shows the magma chamber to be 10 km wide, with its roof at a depth of 8 to 10 km below the surface. Credit: Modified from D. P. Hill *et al.*, *J. Geophys. Res.*, **90**, 11 111–20, 1985.

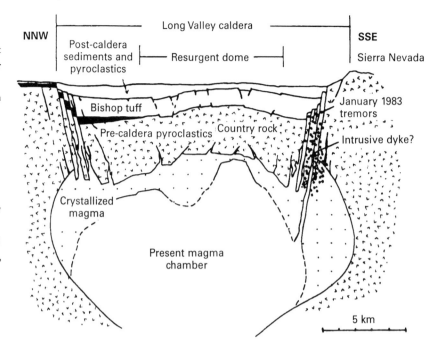

Valles caldera in New Mexico is a showcase example of a resurgent caldera: the basin is 23 km in diameter and has a total drop of only 400 m, with a subtle swell in the center; ring faults with a dozen, small volcanic domes; and vast expanses of ignimbrite fields, deeply furrowed by erosion. Other resurgent calderas have more camouflage, as does the Yellowstone volcanic province, which is 65 km wide and has erupted several times over the past three million years. In contrast to its subdued appearance on satellite imagery, Yellowstone is a magnificent example of resurgent caldera on the ground, with its thick blankets of tuff, and a rich hydrothermal activity of solfataras and geysers.

Only a few years ago, the list of resurgent calderas only boasted several dozen features. But now that they have received wider recognition and satellite attention, resurgent calderas are estimated to number several hundred, disseminated mostly in the continental, back-trench provinces of the Andes, Central America, and the Basin and Range of North America. Mexico alone is estimated to harbor close to one hundred resurgent calderas. In view of their truly colossal eruptions, it is a blessing that resurgent calderas are dormant most of the time: the average interval between eruptions is a million years – the time it takes for the vast magma chamber to recharge with hundreds of cubic kilometers of siliceous, gas-rich material.

The pace of eruptions on Earth

The rhythm of volcanic activity on Earth can be expressed in several ways. First of all, we can state that volcanism on Earth is permanent. Subaerially alone, there are

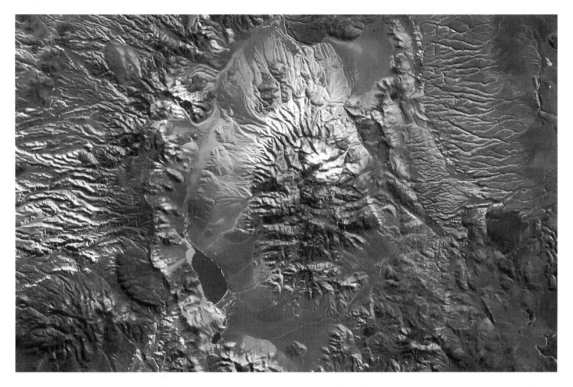

FIG. 2.10. Satellite view of Cerro Galan resurgent caldera in Argentina (26° S, 66° W). The image is a composite using three channels of the Landsat Thematic Mapper. The caldera, in the shape of a pear, is 35 km by 20 km: its central massif is carved by erosion and surrounded by pyroclastic fall-out. A lake is prominent in the southwestern corner of the caldera, as are dark lava domes (notably west of the lake, and at top edge of frame). Resurgent calderas alternate cataclysmic ignimbrite blow-outs and long phases of magma chamber replenishment (accompanied by a doming-up of the central massif). The last major eruption of Cerro Galan took place 2.2 million years ago. Credit: NASA/Landsat,. color-composed by Dr Peter W. Francis at the image processing facility of the Lunar and Planetary Laboratory, Houston, USA. Courtesy of P. W. Francis from his book *Volcanoes of the Central Andes*, Springer-Verlag, 1992.

over 50 volcanoes erupting each year – as reported over the last few decades – and there is no reason to believe that this 'background' volcanic activity has departed significantly from the norm over the centuries.

Eruptions are fairly random in space and time, apart from an occasional 'salvo' when neighboring volcanoes along the same plate margin erupt quasi-simultaneously. One historical example was the eruption of both Soufrière of Saint Vincent, and Mount Pelée of neighboring Martinique, in the Caribbean arc in May of 1902. A more recent example is the setting off of five volcanoes in Papua New Guinea in the year 1974.

Submarine ridges are believed to harbor just as many, if not more eruptions per year than subaerial volcanoes, pushing the number of yearly eruptions on Earth into the hundreds. Estimates of the volume of magma emplacement – both as plutonic bodies and lava flows at the surface – average 15 to 20 km^3 per year along the mid-ocean ridges, which is close to ten times the volume erupted in subaerial settings such as island arcs, hot spot shields, and continental rifts.

Superimposed on this background rate of volcanism are 'spikes' representing infrequent, larger eruptions. Eruptions of the size of Krakatoa ($10\,km^3$ of ash) occur once in a century. Larger, catastrophic eruptions of the resurgent caldera type are rarer yet: estimates call for one ignimbrite blow-out ($1000\,km^3$ or more) every 50 000 years or so: the most recent was the Toba eruption of 75 000 years BP. These giant eruptions have a significant impact on the environment, and are truly catastrophic at the time of their happening, but their contribution to the Earth's crustal output is dwarfed by the more discrete, incessant 'trickle' of background volcanism. Indeed, an eruption of $1000\,km^3$ of magma every 50 000 years boils down to an average figure of $0.02\,km^3$ per year: in the global picture, background volcanism contributes one thousand times more magma than do giant but rare events.

Backing up to a larger perspective, we can look for activity patterns and cycles on timescales of millions of years. There do seem to be variations of volcanic output at this level, linked to changes in plate tectonic motions. In any one province, patterns can be observed: subduction volcanism along the western US cordillera, for instance, rose to an activity maximum around 35 million years BP, followed by a steady decline to a Pliocene 'lull' 5 million years ago, and a new pulse of abundant volcanism over the last 2 million years. Activity cycles on the timescale of a few million years are therefore apparent on a regional scale.

If we further broaden our perspective to contemplate timescales on the order of hundreds of millions of years, a large wavelength pattern can be observed, marked by cycles of huge lava outpours: these are called Very Large Igneous events (or VLI for short). Trap eruptions fall into this category: statistically, there is one trap event – lasting a couple of million years – every 50 million years or so. The latest trap eruption occured in the Columbia River basin around 17 million years ago, and the one before that in the Deccan province of India, 65 million years ago.

These large igneous events are more frequent if we take into account not only the continental, subaerial traps, but the large basalt provinces that occasionally pour out underwater, building giant oceanic plateaux. If we merge, in the same list, continental traps, oceanic plateaux, and ocean opening events, we see that the frequency of very large igneous pulses is more on the order of one mega-event every 25 million years or so. Over the few million years when one is active, the extra lava output averages 1 to $2\,km^3$ per year (the average for the Deccan Traps), and up to 20 or $30\,km^3$ for the most outstanding events (the Ontong Java plateau). This means that the worldwide production of igneous crust rises only slightly during a VLI (by 10%, if we compare $2\,km^3$ to the present $20\,km^3$ output at the ridges), although in extreme cases such as the Ontong Java event, the Earth's magma output does appear to double.

Again, one should stress that these yearly figures are averages, as deduced from volume measurements and dating of entire lava fields. In practice, a $20\,km^3$ magma rate per year derived for one period could just as well mean $10\,km^3$ one year and $30\,km^3$ the next, or even more contrasted figures: the $2\,km^3$ per year of a large igneous event is likely to signify that $1000\,km^3$ is catastrophically erupted in one year, followed by 500 years of quiet until the next eruption. When we discuss volcanic outputs on the Earth and other planets, we must always keep in mind this difference between average rates over time, and the 'instant' output of individual eruptions.

Volcanic regime of planet Earth

To summarize, the Earth's volcanic activity is characterized by several superimposed cycles and trends. First, there is a slow decline of heat production since the Earth aggregated and stratified in the early days of the Solar System. The mantle has cooled close to 200 °C over its history, and volcanic activity is only half as vigorous today as it was three billion years ago.

The Earth's mainstream volcanic activity – primarily at the mid-ocean ridges, with hot spot activity and subduction volcanism contributing an extra 10 or 20% – fluctuates on scales of millions of years, depending on which mid-ocean ridges and hot spots are active at any given time, each readjustment in the boundaries of tectonic plates slowing down or speeding up the global production rate.

Occasionally, every 25 million years on average, a pulse of extraordinary activity takes place, corresponding to the emergence of a super hot spot, or a giant convection loop of magma at the surface. These very large igneous events introduce a distinct, cyclic pattern, although there is some debate as to the exact period of the cycle, and if it carries any meaning. One interpretation is that it might reflect the period of convection patterns in the mantle, perhaps due to lower mantle/outer core interactions 3000 km below the surface. Another explanation is that large asteroid and comet impacts set the pace in some way, by 'ringing' the planet like a bell (indeed, large impacts statistically occur on a 25 million year cycle, as do the very large igneous events).

Whatever its origin may be, the cyclic nature of volcanic activity sets the Earth apart from all the other planets of the Solar System, as we shall see in our final chapter on comparative planetary volcanism.

3 An astronaut's view of the Moon

FIG. 3.1. The Moon viewed from Apollo 16. Credit: NASA.

The mysterious Moon

The Moon is a large rocky body that circles the Earth in a little over 27 days, as it accompanies our planet in its annual revolution about the Sun.

The distance from the Earth to the Moon, and the Moon's size were calculated as early as the third century BC, when astronomers recorded the Moon's position and angular size in the sky, and applied the laws of trigonometry. The first estimates by Aristarchus were remarkably accurate, and needed only slight adjustments throughout history as measuring techniques improved. Today the Moon is known to be 3476 km in diameter, a quarter the size of the Earth. It is 13 times smaller in area, and close to 50 times smaller in volume. The distance between Earth and Moon averaging

384 400 km, we can visualize the pair as having the proportions of a grapefruit and a walnut, placed three meters apart.

The Moon is far enough from the Earth, however, for most of its features to be unresolvable to the naked eye. Only a contrasting pattern of bright and dark patches is usually visible. It was therefore a rather bold speculation for the Greek philosopher Anaxagoras, in the fifth century BC, to postulate that the Moon was a rocky planet with an irregular surface – similar to the Earth – rather than a smooth, divine sphere, as the ruling dogma contended. Anaxagoras was further convinced of the Moon's stony nature after witnessing the fall of a meteorite in his native Greece, which for him was proof enough that rocky worlds existed outside the Earth. Understandably, Anaxagoras was ahead of his time. His revolutionary ideas were pronounced heretical, and the philosopher was sentenced to death (fortunately for Anaxagoras, his sentence was later commuted to life in exile). Thus was the concept of a rocky Moon put to rest for nearly twenty centuries.

Galileo rekindled the controversial idea in January of 1610, when he pointed the world's first telescope at the Moon. The instrument's magnifying power was only 3 ×, but that was enough for the Italian physicist to view the Moon '. . .as having not a smooth and polished surface, but a rough and uneven one, and as is the case with the surface of the Earth, one covered with high elevations and deep hollows. . .'

Galileo was able to observe – as we can today with a pair of binoculars – that the Moon's surface is divided into patches of bright and rough terrane, which he named the *terrae*, and dark, smooth expanses which he initially misinterpreted to be seas, and named the *maria*. As for the bowl-shaped depressions that peppered the bright highlands, Galileo tentatively identified them as craters.

As telescopes improved and magnifying powers increased, craters turned up in great numbers on the Moon. Italian astronomers and cartographers intuitively thought of them as volcanoes, influenced as they were by the local examples of Mount Vesuvius and the Phlegrean craters near Naples. French astronomer Puisieux even went so far as to propose a mechanism for their creation: the lunar craters were collapsed volcanic domes that had blown out all of their gas.

Another origin for the bowl-shaped depressions was put forth in the late 1800s. The bright streaks extending out from the lunar craters looked very much like impact fall-out, leading to the claim that the Moon's formations were collision scars blasted by the impacts of giant meteorites. Indeed, geologists were beginning to identify such impact craters on Earth – the Barringer *Meteor Crater* in Arizona being the most famous.

The very large number of impact craters on the Moon was attributed to the lack of an atmosphere, which allowed most cosmic debris to strike the lunar surface unimpeded, whereas most projectiles headed for the Earth burn up in our dense atmosphere as shooting stars, and rarely make it to the ground. When they do, erosion is quick to erase the impact scars from the surface. On the Moon, by contrast, impact scars would be left to accrue over the years, undisturbed by the lack of significant erosion, and this would explain the landscape's thoroughly battered look.

Volcano supporters did not give up hope, however, and the origin of lunar craters remained a hot issue throughout the first half of the twentieth century. The proponents of meteorite impacts put forth convincing arguments, such as the ballistic nature of rays fanning out from the craters, and the shallow profiles of the bowls, which were more typical of projectiles blasting the surface than of volcanic blow-outs.

Volcanologists answered with their own line of arguments: didn't the bright rays around lunar craters resemble just as closely the streaky fall-out of ash around Aso volcano in Japan? And what to make of that reddish cloud observed during a telescope survey of Alphonsus crater, one winter night in 1958? As space flight drew near, the enigma of lunar craters was still far from solved.

The volcanic nature of the lunar 'seas', on the other hand, was more widely accepted than that of the craters. Making up about a third of the Moon's near side, the dark patches have the low reflectivity and smoothness of basalt, and are reminiscent of the giant lava fields on Earth (flood basalts or *traps*). The fact that lunar lava plains were not associated with any recognizable volcanoes came as no surprise: flood basalts are known to erupt at high discharge rates and low viscosities, which prevent the build-up of steep mounds around erupting vents. The vents themselves become sealed off by late-stage lavas and neighboring flows, which tend to further obscure their exact location. Not to mention that any tell-tale volcanic fissures on the Moon would hardly show up at the coarse resolutions of 500 m, provided by the best telescopes. In order to get a closer look at the lunar landscape, automated spacecraft and eventually man would have to make the voyage.

Voyage to the Moon

With the onset of space travel in the late fifties, the Moon became the target of a grand political race that opposed the Soviet Union and the United States. The true beneficiaries of the Moon race, in the long run, turned out to be the scientists: for the first time in history, man was given an opportunity to explore a planetary body other than the Earth, and draw comparisons with his own world.

The exploration of the Moon was accomplished in three major steps. First came a mapping phase by telescope, while the first lunar probes were being built. Thousands of high-resolution photographs of the Moon's near side were taken at the Pic du Midi observatory in southern France, Lick observatory in California, and Flagstaff observatory in Arizona. The photographs were assembled into detailed base maps that helped geologists unravel the intricacies of lunar stratigraphy – the overlapping sequence of various geological units, such as postulated lava flows and impact ejecta – and allowed mission planners to go about selecting the sites for the upcoming automated and manned lunar landings.

Phase 2 of the lunar conquest involved the automated probes. These light-weight spacecraft had a dual mission: to test the intricacies of translunar space flight, including automated landings; and to gather *in situ* precious information on the environment and surface of the Moon. The Soviet Union headed off the series with Luna 2 in December of 1959, achieving the first impact of a man-made object on the surface of the Moon. This they followed with the very successful Luna 3 mission: the automated probe ducked behind the Moon and radioed back the first pictures of its uncharted far side.

Lunar missions picked up in 1966–1968, starting with the first successful automated landing on the surface of the Moon – the Soviet probe Luna 9 on January 31, 1966. The remarkable Surveyor series, launched by the United States, followed suit with successful landings in the Ocean of Storms (Surveyor 1 and 3); the Sea of Tranquillity (Surveyor 5); and the vicinity of craters Copernicus (Surveyor 6) and Tycho (Surveyor 7). The probes' cameras revealed a compact, dusty and granular soil, not unlike that of a very dry, freshly-plowed field, strewn with angular rocks ranging from a few centimeters to several meters in size (see fig. 3.2).

The Surveyor landers also provided geologists with a crucial new set of data – the first chemical analyses of lunar soil, performed on site by remote-sensing. In these experiments, alpha-ray detectors bombarded the lunar soil with helium nuclei, and measured in return the intensity and wavelength of the backscattered radiation. These early analyses confirmed that lunar minerals, like their earthly counterparts, were composed mainly of oxygen and silicon, with abundant aluminum, calcium, iron, and magnesium, and relatively low concentrations of the alkali metals sodium and potassium.

The Surveyor data was too sketchy to pinpoint which mineral types these atoms made up, but the element list showed a few interesting trends, such as a higher concentration of aluminum (and a correspondingly lower proportion of iron) in the highland terra sites, versus the lowland mare sites. This lunar-wide chemical trend was later confirmed by orbiting spacecraft, using remote-sensing techniques.

In lunar orbit

At the same time that probes were landing at specific sites on the Moon, automated orbiters engaged in a comprehensive survey of the lunar surface, in preparation for the manned missions.

The Soviet Union again was first in mastering the technique, placing Luna 10 in lunar orbit in April 1966, and performing the first gamma-ray survey of the Moon's surface under the spacecraft's flight path. The gamma-ray experiment measured the proportions of radioactive uranium, thorium, and potassium on the surface, showing their concentrations in the mare plains to be typical of basaltic rock.

But it was the American Lunar Orbiters that stole the show, beaming back thousands of low altitude photographs of the lunar surface, with details less than ten meters across. With this deluge of data, the geology of the Moon came into much sharper focus.

First, as was originally revealed by Luna 3, the Moon turned out to be markedly asymmetrical: the far side lacked the vast, smooth maria which were so typical of the near side. Several large basins did show up on the far side imagery – such as Orientale, a 1000 km-wide impact basin with three concentric rings of mountains – but only a few scanty patches of lava were seen to pond in their topographic lows.

FIG. 3.2. In January 1968, Surveyor 7 became the fifth American probe to land automatically on the lunar surface, in preparation for the Apollo missions. The probe landed in the southern highlands, north of impact crater Tycho. The horizon is one kilometer away. Besides collecting thousands of images and proving that the lunar soil – the regolith – was dense enough to support spacecraft, the last three Surveyors analyzed its chemical make-up. Surveyor 7 showed the highland site to contain more aluminum and sodium, and less iron and titanium than the mare sites. Credit: NASA.

Thus it appeared that although both the near side and the far side were battered by massive meteorites during the Moon's early history, only on the near side did the magma upwell to the surface through the cracks, and spread extensively onto the basin floors.

Volcanic features were confined to these near side maria. High resolution imagery revealed smooth, overlapping patches of lava with distinct hues, reflecting subtle differences in mineralogy; long, mysterious ridges buckling the lava plains; and sinuous, meandering rilles, that resembled dried-up river beds. These sinuous channels were especially puzzling, since liquid water was not believed to have ever flowed on the Moon: alternative explanations for the rilles called for giant lava channels, or ash flow gullies of unprecedented magnitude.

Conspicuously lacking, however, were true volcanoes as we knew them on Earth: there were no giant lava shields on the Moon; no steep cones with summit craters; no obvious calderas with collapsed floors and arcuate eruptive fissures. Close-up views of the millions of craters dotting the lunar surface suggested that nearly all of them were impact scars, due to meteorite collisions. Some of the craters did show up occasional volcanic features, such as smooth lava plains nested on their floors, but these were thought to be impact melts – patches of ground melted by the heat of the collision – or later upwellings of lava through the fractures in the shattered crust, rather than proof that the craters themselves were volcanic in origin.

The only landforms that did qualify as potential volcanoes were a handful of low hills on the edges of the maria: the most convincing candidates were an array of low domes in the Ocean of Storms, known as the Marius Hills, and the Rumker and Harbinger domes at the foot of the Aristarchus Plateau. The Marius Hills raised enough interest, in fact, to make it into the select list of primary Apollo landing sites.

Apollo 11: touchdown in a lava field

While the mission planners were busy selecting the landing sites, the Apollo astronauts were undergoing final training in preparation for their lunar missions. A significant part of that training consisted of intensive courses in geology, since the astronauts would be asked to make detailed descriptions of the lunar surface, and collect rock samples to bring back to Earth. One civilian geologist was added to the Astronauts Corps in 1965, but it was a secret to no one that the first men to walk on the Moon would be chosen among the professional pilots, who were in charge of mission operations and oversaw the selection procedure.

The test pilot astronauts turned out to be excellent students in the Earth Sciences: they were well aware that a good command of geology would boost their chances when it came to choosing the final flight crews. In the classroom and out in the field, Armstrong, Conrad, Scott, and their peers learned to recognize rocks and minerals, describe lava flows, and measure the orientations of folds and faults. The astronauts went on field trips with their instructors to Meteor Crater in Arizona and Ries Crater in Germany to study impact features, and to Iceland and Hawaii to examine lava fields. As a result, when Neil Armstrong boarded the Apollo spacecraft in July of 1969, en route for the first lunar landing, his knowledge of geology was close to that of a masters degree.

As it turned out, the Apollo 11 landing site was chosen only for its safe appearance: the primary target was a smooth-looking patch of sparsely cratered plains, on the western edge of the Sea of Tranquillity (see fig. 3.3 for map of Apollo and Luna sites).

After a three-day translunar flight, as they came plummeting over the mare plains in their final approach, Armstrong and Aldrin observed the landing site to be much rougher up close than expected, and performed some last minute maneuvering to steer out of a boulder field. They landed their lunar module (LM) downrange of the

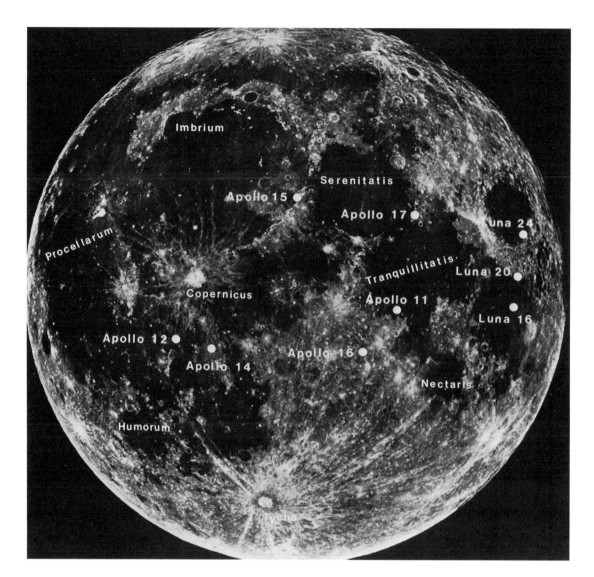

FIG. 3.3. The location of the six Apollo landing sites (and the three Luna sample-return sites) on the near side of the Moon. Credit: NASA photo. From Harrison H. Schmitt, 1991. Evolution of the Moon: Apollo Model, *American Mineralogist*, **76**, 773–84.

target, on a more even surface, in a cloud of mineral and glassy dust kicked up by the engine's exhaust.

As the dust disappeared over the horizon, the two astronauts discovered a landscape of gray, tan and brown rock fragments, embedded in a thick, granular soil – the result of meteorites pounding the ancient lava flows into mineral dust and glass over millions of years. When Neil Armstrong stepped out onto the Moon, his first words – after his historical 'one small step for man' – were to describe this thick mineral blanket: 'The surface is fine and powdery. I can pick it up loosely with my toe. It does adhere in fine layers like powdered charcoal to the sole and sides of my boots. . .'

One of Armstrong's first assignments on the Moon was to collect a spadeful of this soil as a contingency sample – something to bring back in case the moonwalk were prematurely terminated. But the walk went on according to plan, and Armstrong was soon joined by Buzz Aldrin. As both astronauts went on collecting samples around the LM, Aldrin described the sparkling, pristine quality of the lunar rocks: minerals jumped out of their matrix with a glistening luster that reminded the astronaut of shiny mica flakes. As it turned out, the sparkling was caused not by mica – mica is a water-bearing mineral and there is no water on the Moon – but by glass, and embedded pyroxene, and feldspar crystals.

Based on their overall appearance, Armstrong guessed the rocks to be basalts. Many were vesicular, riddled with small, bubble-like holes, which the astronaut ascribed to the permanent striking of the rock surface by micrometeorites. The vesicles, in fact, turned out to be the frozen shapes of gas bubbles that formed in the magma, as the molten rock flowed and cooled upon the lunar surface.

Throughout the two-hour moonwalk, the astronauts collected a variety of vesicular samples; several fuller and denser chunks of lava; many impact breccias; and spadefuls of lunar soil and pebble-sized fragments, amounting to a grand total of 21 kg of 'lunar luggage' to bring back to Earth.

The lunar receiving laboratory

Four days later, the two happy moonwalkers and Michael Collins, their orbital crewmate, were warmly welcomed to the Lunar Receiving Laboratory in Houston, along with their historic collection of rock specimens from another world. The astronauts and their samples were quarantined in the specially designed complex – in the improbable event that they had been exposed to some microscopic form of alien life on the Moon.

Selected rock samples were immediately subjected to a series of in-house preliminary tests, and the rest of the collection partitioned into two batches: samples to be shipped out for analysis to the national and international scientific community; and samples to be stored for the benefit of future generations.

There are three kinds of analyses performed on rocks that provide precious clues as to their nature and history. The first method is hand specimen observation, followed by the analysis of a thin, translucent slice of the rock under a polarizing microscope: by studying the crystals, one can estimate the rock's mineral make-up, and whatever history of shocking or reheating it underwent.

A second type of analysis is to bombard a rock sample with an electron beam, and observe the X-rays emitted in return by the excited constituent atoms. This X-ray spectroscopy gives a precise reading of the proportions of various atoms that make up the rock.

In a third type of experiment, in order to probe even farther into the sample's atomic make-up, a fragment of rock is pulverized into gas and run through a mass spectrometer: different varieties of each atom are sorted out according to their isotopic mass, and the readings provide valuable geochemical information, such as the rock's age (the technique of radiochronology).

From the preliminary tests in Houston, it was confirmed that many of the Apollo

11 moon rocks were volcanic indeed. Under the microscope they showed mainly pyroxene and feldspar crystals; with a sprinkling of iron and titanium oxides; and traces of olivine in some samples, and of quartz in others. This mineral make-up placed the mare rocks in the basalt category.

Upon further examination, with the X-ray fluorescence technique, lunar basalts revealed a distinct chemical personality, that set them aside from terrestrial basalts. In particular, Apollo 11 basalts were found to contain ten times more titanium oxide than their earthly counterparts (12%, versus 1 to 2%). Titanium is a refractory element – a metal with a high melting temperature. Other atoms of the refractory family were also found to be significantly enriched in the lunar rocks. On the other hand, volatile elements – elements that melt at modest temperatures, such as sodium and potassium – were found to be depleted: the Apollo 11 samples displayed five times less potassium and sodium oxides than comparable terrestrial lavas. The most volatile of oxides – hydrogen oxide, otherwise known as water – was even completely lacking in the Apollo samples: as we noted earlier, lunar minerals are absolutely waterless.

The most exciting piece of data to come out of the isotopic analyses was the age of the lunar basalts – how long ago the lavas were erupted onto the lunar surface. A conservative consensus had emerged prior to the Apollo landings, suggesting that the mare basalts would have to be very old indeed – perhaps as old as three or four billion years – since the Moon was such a small planetary body that it could not possibly have stayed warm and fueled volcanic activity for very long. The number of impact craters accumulated on the lunar surface supported the claim of very ancient flows, undisturbed for billions of years.

Challenging this interpretation were a few geologists who supported the idea of some limited volcanic activity extending much later in time. These 'volcanic optimists' were hoping that some of the samples would turn out to be less than a billion years old.

The 'conservatives' won the first round. Isotope analyses of the Apollo 11 samples, using both the rubidium-strontium and the potassium-argon methods, yielded ages clustered around 3.7 billion years: the first rocks brought back by the astronauts were as old as the oldest rocks on Earth. The Moon was truly a showcase of the earliest forms of volcanism in the Solar System – a priceless planetary museum.

The Ocean of Storms

Four months after the triumphant Apollo 11 mission, a second crew headed out for the volcanic maria of the Moon's near side. Aboard Apollo 12's lunar module *Intrepid*, astronauts Pete Conrad and Alan Bean made their landfall on November 19, 1969, in the southern stretches of the Ocean of Storms, 300 km south of Copernicus crater.

The Ocean of Storms occupies most of the western half of the Moon's near side: it is the largest patch of dark mare visible to the naked eye, with a somewhat irregular shoreline. Apparently, the vast Ocean of Storms was a thin set of lava sheets spreading across shallow, interconnecting basins, in contrast to the deeper lava piles of the downwarped, more circular basins.

The Apollo 11 samples had shown that the lunar maria were giant basalt flows, that had flooded the impact basins on the Moon's near side. But many questions remained unanswered: did all the basalts come from one unique, chemically uniform magma batch inside the Moon, or did the lavas embrace a wide range of chemical compositions, pointing to different magma sources? Were all lunar lavas of roughly the same age? Did they shoot up from the depths of the Moon in response to the shock waves of the giant meteorite impacts that carved out the basins, or did they erupt 'on their own', independently of the basin-forming events? Apollo 12 provided an important set of new data, that helped address these issues.

As a whole, *Intrepid's* landing site resembled Apollo 11's Tranquillity Base: it was a monotonous rolling plain of lava boulders set in a thick bed of mineral pebbles, dust, and glass. Impact craters, ranging in size from a few meters to a few hundred meters in size, with smooth, rounded profiles, gave to the plains the battered appearance of an abandoned artillery range.

Whereas their Apollo 11 predecessors had spent little more than two hours on the lunar surface – and strayed no farther than 50 m from the LM – astronauts Conrad and Bean performed two long moonwalks lasting four hours each, taking them nearly half a kilometer from the lunar module (see fig. 3.4). Besides setting up a new science station, their main objective was to collect a wide array of lunar rock and soil samples, and to explore the slopes of a close-by impact crater, roughly the size of a football field.

Lunar flood basalts

During their moonwalks, Conrad and Bean picked up 69 samples amounting to a grand total of 34 kg – nearly twice the amount collected by the Apollo 11 astronauts. Upon reaching Houston, these new samples underscored the great diversity of the lunar lavas.

The Apollo 12 samples were found again to be basalts, and fell into two categories: quartz basalts, with large pyroxene crystals visible to the naked eye; and olivine basalts, sparkling with specks of the pale green mineral. As a bonus, a third category of basalts was represented by one peculiar, very aluminous sample.

Such a variety of lava compositions and textures is remarkable: the spread is substantially greater than one would expect from a similar field trip on Earth. Moreover, the Apollo 12 samples are noticeably different from the Apollo 11 rocks: they bear much less titanium oxide (3% instead of 12%); although, on the other hand, they share the same richness in iron and magnesium (20% and 10%) – which is typical of the lunar lavas, and is one reason for the great fluidity of the flows (magmas that are rich in iron and magnesium are high-temperature melts, and are naturally very fluid).

The fluidity of the lunar lavas can be measured directly, by melting, in a high-temperature oven, chunks of the samples brought back by the astronauts. The experiment shows that lunar basalt fully melts at a temperature of 1200 °C, yielding a

FIG. 3.4. Apollo 12 demonstrated precision landing on the Moon by reaching a target site in the Ocean of Storms, a few hundred meters from the automatic Surveyor 3 probe that had landed two years prior. A visit to the probe was scheduled during the astronauts' second moonwalk. Alan Bean inspects the spacecraft, photographed by Pete Conrad. The Lunar Module is visible on the horizon. Credit: NASA.

magma of very low viscosity: 5 to 10 poise, which is roughly the viscosity of motor oil at room temperature. By comparison, basaltic magmas on Earth display much higher viscosities – a few hundred to a few thousand poise, as measured during recent eruptions, although prehistorical basalts of the *trap* type were probably erupted at viscosities as low as 30 poise, much closer to the lunar figures.

The Apollo 12 site is traversed by at least two different lava flows – or so it appears on photographs taken from lunar orbit. The flows have distinct albedos, and slightly different densities of impact craters, suggesting different ages (the most cratered flow being the oldest). Apparently, astronauts Conrad and Bean sampled the older flow, which turned out in the lab to be 3.30 billion years old, with one sample coming in at 3.16. On average, the Apollo 12 basalts were thus 500 million years younger than their Apollo 11 counterparts (3.7 by).

This difference in age was a fundamental finding, in that it proved that lunar basalts were indeed very ancient, but that they spanned a substantial interval of time, stretching much beyond the period of heavy bombardment that blasted out the great basins, and which ended 3.8 billion years ago. The relative youth of the Apollo 12 samples rekindled the hope of finding even younger volcanic lava on the Moon, starting with the less cratered flow that came close to the Apollo 12 site, but was out of range of sampling. Judging from its lower density of impact craters, that flow was likely to be about 500 million years younger than the samples dated at 3.3 billion years – i.e. closer to 2.7 billion years 'only'. And by counting impact craters on other mare imagery, it appeared that some lunar flows were younger still.

Another challenging task was determining the total volume of these lava flows, that poured layer after layer onto the lunar surface. Measuring the areal extent of the flows was easy, but measuring their aggregate thickness was quite another matter. Were lunar lavas a thin veneer of rock only a few meters deep, or were they great pile-ups reaching thousands of meters in cross section?

For clues, we can turn to the great lava fields on Earth – the basalt traps that most closely resemble the lunar maria. Traps like those of the Columbia River Plateau, and India's Deccan region, consist of many dozens of thin flows – each 10 to 20 m thick – that add up to total thicknesses of 1000 to 2000 m. Their structure is most evident in places where rivers (like the Columbia River in the western US) cut through the volcanic pile and expose the layers in cross section.

Unfortunately, there are no rivers on the Moon to provide such a perspective, but geologists can use impact craters instead: old craters surrounded or flooded by younger lavas give a rough idea of the flow thickness, based on how much their rims still protrude above the lava. Inversely, impact craters occurring after the emplacement of lava flows also provide valuable insight by blasting and overturning the deeper layers onto the surface as ejecta rays. The Apollo 12 landing site was chosen on an ejecta ray of the great Copernicus crater precisely for that reason, and indeed – as we shall see in the next chapter – some peculiar material from below the volcanic pile did turn up in the rock samples returned by the astronauts.

Cross section of the lunar maria

A good way to get a precise idea of the thickness of mare basalts is to measure seismic waves as they are sent travelling through the rock layers by moonquakes and meteorite impacts. Seismometers set up by the Apollo astronauts picked up enough signals on several landing sites to provide detailed profiles of the maria structure.

The topmost layer invariably consists of a veneer of mineral rubble – fragments broken up and mixed together by billions of years of meteorite bombardment. This lunar 'soil' is named the *regolith*, and averages 5 to 10 m deep in the mare plains. Beneath the regolith lie the more compact lava layers, also somewhat broken up and fractured by the meteorite pounding. On the Apollo 12 site, the lava layers average 200 m deep, down to the basement floor. In other areas of the Ocean of Storms, the volcanic pile is estimated to be over 1000 m thick, and in the more circular, deeper basins, volcanic flows are thought to have accumulated in some places to aggregate thicknesses of over 6000 m.

It is possible to estimate the individual thickness of a single lava flow within a volcanic pile. This can be achieved by measuring the mineral grain size in lava samples, which is indicative of the cooling rate of the lava flow – and therefore its thickness. The small grain size of the Apollo basalts indicate that their flows are no more than 10 m thick, comparable to individual trap flows on Earth.

From the samples and observations of the first two Apollo landings, one could therefore get a clear picture of the composition, age, and extent of the lunar lavas, but information on where and how actual eruptions took place was still lacking. Did the magma upwell through fissure vents in fiery fountains high above the lunar surface? Were there occasional blow-outs of gaseous pyroclasts – surges of hot ash rolling down the lunar slopes?

The next two Apollo missions did little to advance the issues. Apollo 13 was an aborted mission that did not land on the Moon. Apollo 14 explored the Fra Mauro region of the Ocean of Storms, which turned out to be hills of impact breccia with very little in the way of volcanic specimens: only 9 out of the 97 samples returned were diagnosed as crystalline lava.

Fortunately, much insight into eruptive mechanisms on the Moon was provided by the Apollo 15 mission, which set out on July 27, 1971 for the eastern edge of the Sea of Rains, and a meandering channel known as Hadley Rille.

Excursion to Hadley Rille

Hadley is one of many rilles that cross the lunar maria. Rilles average between 10 and 100 km in length, with a record 340 km for the longest one; and range from a few meters to 3000 m in width.

There are two types of rilles: linear rilles, and sinuous rilles. Linear rilles often cut across preexisting topography, which shows them to be fault-related: they are close cousins to our *graben* faults on Earth, and most likely result from extensional tectonics, that stretch the crust and down-drop swaths of terrane between pairs of facing fault planes.

Sinuous rilles, on the other hand, are much more reminiscent of meandering river beds than they are of tectonic cracks. There are several hundreds of these enigmatic valleys on the Moon, that wind across volcanic plateaus, or snake along the edge of the great mare basins, as does Hadley Rille. Many can be traced back to sources along fracture rings, mountain clefts, and other basin-edge structural trends.

From the vantage point of lunar orbit, Hadley Rille is seen to emerge from an arcuate cleft in the Apennine mountains, and descend to the floor of the Sea of Rains, where it runs northward nearly 100 km along the edge of the basin. The rille averages 1200 m in width, with a V-shaped cross-section, and a rim-to-bottom depth of 300 to 400 m along most of its course.

Hadley's bends are typical of flow and erosion along a channel, which led to the early speculation that the rille might have been carved by running water. This theory was quickly dispelled as it became evident that there was no water in the chemical make-up of the Moon. Even if there had ever been, the release of water in a running, liquid form would have been inhibited by the near-zero vacuum of the lunar environment. A much stronger hypothesis was that Hadley Rille, like other sinuous channels of its type, was volcanic in nature and resulted from massive outpourings of fluid magma, that channeled their way downslope along favored, structural lines.

It was with this volcanic model in mind that astronauts Dave Scott and Jim Irwin flew their LM over the Apennine mountains and down onto the lava plains, landing about a kilometer east of Hadley Rille's right bank (see figs. 3.5 and 3.6). Their mission was complex: for the first time in the Apollo program, the astronauts brought along an electric-powered lunar rover, folded up against the LM's descent stage. Less than 200 kg in weight, this four-wheel drive vehicle allowed the astronauts to rove about the surface at an average speed of 10 km/h, and carried a TV camera, sampling tools, and casings for the collected rocks. Over the course of three outings – each lasting over 6 hours – astronauts Scott and Irwin drove 28 km around the landing site, making stops at ten major locations to take photographs, drill cores, and collect a grand total of 370 rock and soil samples – a record hoist of 77 kg.

The first outing was scheduled to take the astronauts to the edge of Hadley Rille. In the vicinity of the landing site, the channel hugs the base of the Apennine mountains before taking a sharp turn at Elbow crater, and striking northwest across the lava plains. At the end of this linear segment, the rille reconnects with the mountain front and resumes its winding course along its base.

A dried-up lava bed

On their first outing Scott and Irwin drove due south, to Elbow crater and the bend leading into the straight portion of the rille. Along the way, the astronauts were struck by the awesome beauty of the mountain range unfolding to their left. The lunar Apennines were not traditional mountains in their earthly sense: they were not layers of rock folded into tight waves by compressive forces acting horizontally – as is the case in terrestrial 'plate tectonic' ranges – but fault blocks uplifted by the great impacts that carved out the lunar basins.

Specifically, the lunar Apennines were formed by the Imbrium impact, which uplifted arcuate mountain blocks along concentric faults, all around the basin. One such block is Mount Hadley, which towers 4500 m above the Apollo 15 landing site. To the south, where the astronauts headed off on their first 'moondrive', neighboring Hadley Delta peaks 3500 m above the basin. As they drove toward the mountain front, Scott and Irwin noticed subhorizontal layers cutting across the slopes, presumably the sheets of ejecta that made up the uplifted blocks, interlayered with igneous intrusions.

At the base of Hadley Delta, the astronauts' main objective was the exploration of Hadley Rille, which hugged the foothills before striking out into the plains. As they parked their rover on the edge of the rille near Elbow crater, Scott and Irwin discovered sloping walls 25 to 30 degrees steep, leveling off to a gentler, talus-like lower section (see fig. 3.7).

Broken-off boulders of all shapes and sizes – some as big as houses – littered the channel bed, 350 m below the rim. In the steep upper walls, compact rock layers protruded through the rubble like so many ribs, manifestly the source of the loose boulders below. Scott and Irwin were well positioned on the ledge of the rille to

FIG. 3.5. The downstream section of Hadley Rille, as it hugs the Apennine mountain front (near St George impact crater) before veering northwest across the lava plains to the foot of another mountain scarp (upper right). The view is toward the west (north is to the right). The rille is an average of 1200 m wide and 300 m deep: in this orbital image, half of its course is visible, spanning approximately 40 km. The Apollo 15 landing site is in the bottom part of the frame, on the eastern bank of the rille between two clusters of small craters (see map, fig. 3.6). The rightmost cluster of craters – the North Complex – appears to have some relief and was believed before the mission to have a volcanic origin. Credit: NASA.

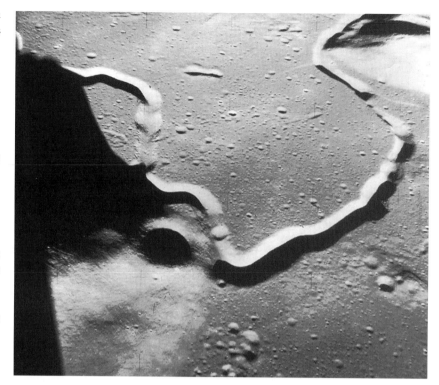

photograph the layers protruding from the facing bank. They took pairs of stereo images, stepping sideways between shots to change the viewing angle; as well as close-ups of the protruding rock with their 500 mm telephoto lens (see fig. 3.8).

The bedrock sticking out from the rille's upper slope showed up to be composed of two separate layers: a dark flow at the top, two meters thick, capping a massive layer twenty meters thick. A few more layers of bedrock, finely stratified, stuck out in places from the rubble farther downslope; and below that, talus deposits completely covered up the bottom half of the rille walls.

The astronauts managed to collect loose fragments of the bedrock by hiking down the slope of the rille and sampling the talus deposits. Back in the lab, these samples turned out to be ferromagnesian basalts, similar to the lavas of Apollo 11 and 12. Again, there was no trace of water in the samples; nor were there any pyroclastic fragments that would hint to a surge flow or *nuée ardente* as the rille's carving agent. As originally suspected, Hadley Rille was rather a giant lava channel that once carried hot magma from an eruptive vent in the Apennine foothills, and delivered it downstream to the flow fronts in the Sea of Rains.

FIG. 3.6. Map of the Apollo 15 landing site, on the eastern bank of Hadley Rille (corresponding to the lower half of fig. 3.5). Radiating from the LM target point are the itineraries of the three outings accomplished by astronauts Dave Scott and Jim Irwin with the Lunar Rover. The first loop to the south allowed sampling of the Apennine front and an axial view of Hadley Rille (see fig. 3.7). The last loop to the north allowed for stops on the edge of the rille (the 'Terrace', see fig. 3.8), and a visit to North Complex. Credit: NASA/Defense Mapping Agency.

FIG. 3.7. Astronaut and Lunar Rover on the edge of Hadley Rille. taken from Elbow crater, the view looks north along the axis of the channel. In the distance (about 10 km away), the channel takes a bend to the left. The foothills of the Apennine mountains loom on the horizon. In the foreground, the rille is nearly 1500 m wide and 300 m deep. Credit: NASA, courtesy of the National Space Science Data Center, through the World Data Center-A for Rockets and Satellites (Apollo image suppliers: F. J. Doyle, R. J. Allenby and F. El-Baz).

Lava channels and lava tubes

Lava channels of such large size do not occur on Earth, although smaller features are common place: volcanologists are familiar with lava tubes and open channels winding their way down the slopes of basaltic shields, as in Hawaii, La Réunion, or Mount Etna, to name but a few sites. These terrestrial channels average a few kilometers in length – seldom more than ten – and are rarely over fifty meters wide.

On the Moon, it would appear that the superior dimensions reached by lava channels are caused by the extraordinary discharge rates of magma out of the eruptive vents. Hadley Rille most likely channeled close to a thousand cubic meters of magma per second down its course, a discharge rate comparable to that of the River Thames in London, or the Charles River in Boston. The fiery flow probably displayed the turbulence of a mountain torrent, rather than the laminar flow of a stable river, a conclusion reached after some theoretical modeling of the behavior of low viscosity magma flows.

The Hadley Rille mission, in this respect, brought a new perspective to the study of volcanic flows, particularly the formation and evolution of lava tubes and channels. In the wake of the Apollo 15 mission, geologists explored anew the lava tubes of the

FIG. 3.8. West wall of Hadley Rille, viewed with a telephoto lens (opposite the 'Terrace' stop, see map, fig. 3.6). The steep wall shows distinct lava layers protruding through the talus rubble. The top bright unit is approximately 20 m thick, with joints dipping to the right. Below it is dark talus and a thinner bright outcrop, 5 to 10 m thick, with distinct horizontal layering. The outcrops are interpreted to be lava flows that took turns flowing down the channel. In the lower part of the image, a jumble of bright boulders protrudes from the talus. Credit: NASA, courtesy of the National Space Science Data Center, through the World Data Center-A for Rockets and Satellites (Apollo image suppliers: F. J. Doyle, R. J. Allenby and F. El-Baz).

Hawaiian volcanoes, the Cascades range, and the Snake River Plain of Idaho, drawing comparisons and testing models. Of particular interest was the contradictory nature of lava channels – both constructive and erosive. Did lava channels build up their beds in a constructive fashion, lining their course with layer after layer of chilled lava each time an eruption took place? Or did the rushing magma cut through the basement rock with strong erosive power, melting and plucking away an ever deepening channel bed?

Piecing together the data from both Earth and Moon, geologists found evidence for both types of behaviour in lava channels: constructive and destructive.

For one, a steady, average-rate discharge of magma will occur as laminar flow, promoting thermal stratification in the magma: the top and bottom of the flow will cool faster than the middle section, because heat is radiated or convected away at the top (vacuum on the Moon, air on Earth), and conducted away at the bottom by the cold basement rock. As a result, construction will occur: chilling at the bottom will gradually build up a higher bed. Likewise, the top of the flow will chill and crust over to form a solid 'roof' of basalt arching over the channel. Ultimately, the channel will seal itself off into a rocky tunnel, or lava tube. Because of the excellent insulating capacity of solid rock, the magma flowing in a closed tube will retain most of its heat, and flow swiftly over large distances. Hence, lava tubes are efficient feeder lines between eruptive vents and lava fronts downstream: they contribute significantly to the construction of volcanic shields and lava plains on Earth.

On the other hand, when the magma discharge rate at the eruptive vent is exceptionally high, and the magma very hot and fluid, flow behavior changes. Rushing into topographic lows – preexistent gullies or fault lines – the confined magma no longer flows as laminar sheets, but rolls and mixes turbulently, and the thorough thermal mixing prevents chilled floors and roofs from forming at the bottom and top of the flow. On the contrary, energy – both thermal and dynamic – is transmitted to the channel bed: the churning lava melts and plucks away chunks of basement rock, which are carried off in the turbulent flow, and assimilated in the melt. Through this process of thermal erosion, the channel digs itself ever more deeply into the underlying strata.

The origin of Hadley Rille

Be they laminar or turbulent, lava rivers are guided in their development by the regional slope; the nature of the substratum they flow upon; and favored structural directions, such as fault lines.

Structural control of channel development is well documented. On Earth, many rivers follow fault lines, as do the few large lava channels that have come under recent scrutiny. The Giant Crater Lava Tube of California follows a graben over most of its 28 km course: it dug its channel in the soft riverbed gravel lining the fault-bounded corridor.

On the Moon, Hadley Rille is also seen to strike along a regional trend, underlined by northeast and northwest-trending faults. From the vantage point of lunar orbit, it was clear to the Apollo 15 astronauts that the channel originated in the elongate cleft at the foot of the Apennine mountains, and that the cleft was itself part of a major lineament extending much deeper into the mountain range. In the lower portion of Hadley's course, the channel also follows major structural trends: many of its straight segments line up with nearby faults – tectonic directions which are clearly related to the Imbrium impact, and are either radial or circumferential to the giant basin.

On the surface, Scott and Irwin were also impressed by the rille's linear trend in the vicinity of the landing site, summarizing in their mission report that their 'first impression of Hadley Rille is that it results from a fracture in the mare crust'.

Lava at Hadley thus appears to have borrowed preexisting graben in the fractured basement rock to dig its rille. Vast quantities of lava were funneled in this way down into the basin – perhaps as much as 5000 to 9000 km^3 – ultimately spreading out to coat the basement rock on either side of the rille's course.

When lava flows down a channel, it does not always stay confined. Occasionally, swells in the discharge rate cause the lava to surge out of the channel bed, and spread into the surrounding plains. When the magma is very fluid, the surge can be spectacular, with waves of molten rock climbing up the banks and splashing over obstacles, leaving a glistening veneer of chilled lava over the landscape, as the flow ultimately ebbs back into its channel. On Hawaiian volcanoes, such 'high tide' bench marks are found up to 10 m above the base level of some channels.

At Hadley Rille, Scott and Irwin found several indications of surge-type behavior. Near Elbow crater, where the rille takes a sharp bend to the left, the astronauts exploring the right bank stood 30 m higher than the left bank facing them across the channel. One interpretation was that the molten magma overflowed to the right as it gushed into the left-turning bend, under the influence of its centrifugal force, building up a 'plastering' of basalt on the outer bank. However, the elevation difference between the facing banks could also result from normal fault motion along the rille – and the down-dropping of the western block – which would be further suggestion of the rille's structural control.

Another observation made by Scott and Irwin was that the Apennine foothills close to the lava channel displayed a dark bench mark 80 m above plains level. The foothills are a few kilometers from the rille, but the bench might still be a 'high tide' mark caused by a dramatic surge of magma leaping out of the channel and flooding the plains. When the discharge rate returned to normal, the overspill would have drained back into the channel.

Another, less dramatic explanation is that the 80 m offset did not result from a lava flood rising along the mountain base, but was caused instead by an opposite motion: at the termination of volcanic activity, the entire lava plain flexed under its own weight, slipping 80 m down the mountain scarp. In this model, the bench mark on the mountain face represents the original level of the lava plain, before it foundered.

Volcanic history of the Hadley site

Piecing together the observations made by the astronauts and the analyses of rock samples they brought back, geologists came up with a detailed history of the Hadley site.

Volcanic events were ancient, as they were found to be on previous Apollo sites: the basalts collected along the rille and in the surrounding plains ranged in age from 3.3 to 3.4 billion years. They were predominantly pyroxene basalts, with an olivine variety collected from the thin dark layer at the top of the rille. The bulk of the lava discharged at Hadley Rille had therefore been the pyroxene basalt, except for a last batch of olivine-rich lava closing the eruptive cycle and capping the sequence.

Away from the rille, Scott and Irwin sampled rocks shed off the mountain slopes: these were found to be mostly shattered crustal rock, broken up and mixed together by the giant impacts that marked the Moon's early history. But the astronauts also found a third type of basalt, much older than the lavas of Hadley Rille (3.85 by). Apparently, these fragmented, aluminous basalts lined the basement of the Imbrium basin: today, they only crop out on the edge of the basin, and wherever meteorite impacts have unearthed them from under the more recent, Hadley Rille flows.

Last but not least, the astronauts sampled a sparkling, bead-rich soil, that was later resolved under the microscope into myriads of tiny droplets of volcanic glass, some of a beautiful emerald green color. This glass was found to be magma spray – chilled droplets of lava that shot up from the vents on the site of the Hadley eruptions, much in the style of hawaiian eruptions on Earth.

In summary, the following history was pieced together for the Hadley region :

1) A little over 3.85 billion years ago, a gigantic impact excavated the Imbrium basin. The blast splashed the Moon with breccia and impact melt, uplifted mountain rings around the basin, and shattered the local crust with both radial and circumferential faults (formation of the Apennine mountains, including Mount Hadley).

2) Following the impact, aluminous magma upwelled through the cracks and flooded the basin, lining it with a first layer of basalt – a layer that is now mostly buried under the more recent flows. It is represented in the sample collection by the aluminous basalt recovered along the mountain front, and dated back at 3.85 billion years.

3) Then came a long, quiescent period lasting 500 million years, when nothing much happened besides the usual hammering of meteorites, and occasional moonquakes and landslides. The aluminous basalt layer was progressively and thoroughly shattered by the impacts, and the battered surface disrupted by normal faulting, especially along the mountain/basin boundary, where long, deep grabens developed.

4) A major phase of volcanic activity flared up 3.3 billion years ago throughout the Imbrium basin. In the Apollo 15 landing area, magma discharged at a tremendous rate from a cleft in the Apennine foothills, and dug its channel along the preexising graben, giving birth to Hadley Rille. Rich in iron and magnesium, the magma overflowed into the plains, and solidified into great sheets of pyroxene-rich basalt. The last dregs of magma that came rushing down the channel turned out to be chemically distinct, and capped the bulk of the flows with a thin veneer of olivine basalt.

5) Locally, pyroclastic fire fountains played along fissures and sprayed the surface with green glass. But after these last eruptions took place, volcanic activity shut down forever in the Hadley region. The lava plains cooled, densified, and progressively settled along the basin's bounding faults. Only a rare meteorite impact shattering the lava plains, or a dislodged chunk of bedrock tumbling down the mountain slopes were to disturb the scene over the aeons to come. Three billion years were to pass before two creatures from another world landed at Hadley Rille, and undertook to unravel its history.

Looking for volcanoes

Apollo 15 was a highly successful expedition, especially for planetary geologists and volcanologists. But despite the excitement and the extraordinary results, the Apollo program was coming to a close.

In the Spring of 1972, Apollo 16 set off for a visit to the lunar highlands, and the jumbled terrae around Descartes crater. Astronauts Young and Duke discovered a landscape of shattered crustal rock, with little evidence of 'continental' volcanism: the suspicious-looking pods of smooth terrane stretching between craters turned out to be ejecta debris dating back to the great basin collisions, rather than volcanic ash.

And then came the shattering news: Apollo 17, scheduled for later that year, would be the last lunar mission. Under the pinch of budget constraints, NASA was forced to wrap up the most exciting exploration program in the history of mankind. Planetary geologists were faced with an excruciating choice: which site on the Moon should they pick for the ultimate mission?

Two schools battled over the options: on one hand, volcanologists wanted a last shot at finding young volcanic flows on the Moon, to prove that some igneous activity had persisted much later than was commonly thought. At the opposite end of the spectrum, planetary geologists who were attempting to piece together the early history of the Solar System wanted access to the oldest rock layers that the astronauts could reach.

In the end, both clans were satisfied: the selection committee picked the valley of Taurus-Littrow, a young-looking basaltic plain nested between the uplifted blocks of an ancient mountain range, on the eastern margin of the Sea of Serenity (see fig. 3.9). During the Apollo 15 mission, astronaut Al Worden had conducted a thorough survey of the area from lunar orbit – while waiting for his companions to return from Hadley Rille – and taken many high-resolution photographs of the valley. Of great interest in the photographs were a prominent landslide scarring the southern mountain face, that would give the roving astronauts an opportunity to sample blocks of primitive material shed off the slopes; as well as a sprinkling of very dark craters and associated deposits on the valley floor, that looked like relatively recent volcanic vents surrounded by their ash fall-out.

A geologist on the Moon

Apollo 17 was the last flight to the Moon, but it was also the occasion of a great premiere: bending to pressure from many interests, NASA's Astronaut Office assigned a civilian scientist to the last Apollo crew. Chosen to ride with Navy astronauts Gene Cernan and Ron Evans, the lucky and deserving scientist was Harrison 'Jack' Schmitt, who held a doctorate in geology.

Schmitt had joined the Astronauts Corps in 1965, and spent the following seven years in training, learning to fly fighter jets – he had never flown a plane before – and running the full gamut of Apollo mission rehearsals. His geological expertise was impressive: Schmitt held a bachelors degree in the Earth Sciences from Caltech, worked on his thesis in Norway, and obtained his doctorate from Harvard, before joining the US Geological Survey in Flagstaff, Arizona, and specializing in the mapping and photo-interpretation of the lunar surface.

Acting as a flag-bearer for the entire scientific community, Jack Schmitt took his berth alongside Gene Cernan and Ron Evans aboard the Apollo 17 spacecraft on December 6, 1973, and the three astronauts blasted off on Man's last voyage to the Moon. On December 11, at 19 h 55 mGMT, Cernan and Schmitt landed their LM in

FIG. 3.9. View to the west from the window of Apollo 17's Lunar Module, looking down onto the dark lava plains of Taurus-Littrow (compare with map, fig. 3.10). The landing site is at the far end of the valley, just before the light-colored tongue of avalanche deposits extending onto the floor from the bright massif in center frame. Taurus-Littrow was chosen as the last Apollo landing site on account of its juxtaposition of both old terrain (the mountains) and apparently young volcanic features on the valley floor. Credit: NASA.

the Taurus-Littrow valley, less than two hundred meters from their target (see map, fig. 3.10).

The landscape was breathtaking: the astronauts beheld a flat, boulder-strewn valley, fifty kilometers long and seven kilometers wide, boxed in by lofty mountain ranges that towered 2000 m above the valley floor. Detached from the slopes, large boulders could be traced down to their resting place, leaving conspicuous grooves in their wake. One particularly large avalanche apron, southwest of the landing site, was scheduled for sampling on the astronauts' second traverse.

Before they could devote their time to rock sampling, the two astronauts began by installing the remote-sensing station: a gravimeter; a particle detector; a seismometer to record moonquakes; and a thermocouple to measure the heat rising from the depths of the Moon. Installing the instruments took longer than planned: as soon as they were finished, Cernan and Schmitt boarded the rover for a quick excursion across the valley floor to the east, and a sampling stop at Steno crater – a fresh impact structure.

FIG. 3.10. The Apollo 17 landing site in Taurus-Littrow valley (north is to the right): compare with regional photo, fig. 3.9. The valley is encased between mountain blocks uplifted by the Serenitatis impact, and flooded by lava flows. Astronauts Cernan and Schmitt accomplished three traverses with the Lunar Rover: to Steno impact crater (1); to the light mantle deposit at the foot of South Massif, with stops on the way back at two dark-halo craters thought to be volcanic (2); and to North Massif and its boulder fields (3). Credit: NASA/USGS base map by D. H. Scott, B. K. Lucchitta and M. H. Carr.

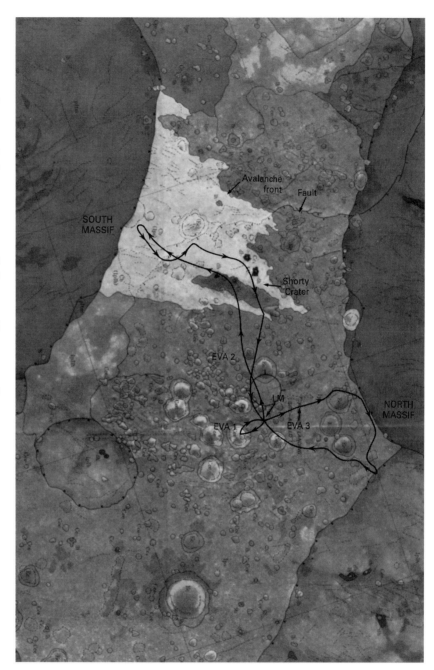

Schmitt later noted in his mission report that, unlike fieldtrips on Earth, his lunar traverses were affected by this permanent quest for speed and efficiency, that prevented him from consciously memorizing his visual impressions. It was only much later, during debriefing sessions back on Earth, that the astronaut was able to retrieve many fundamental observations stored in his subconscious.

Despite this handicap, Schmitt was able to appreciate the exceptional conditions offered by the lunar environment: with no atmosphere to diffuse the sunlight, rock surfaces were ideally lit, crystals glistened, and differences in texture were prominently displayed. Only a light, brownish patina veiled those rock surfaces most harshly exposed to sunlight and cosmic rays. During their first sampling session, Cernan and Schmitt collected fragments of bedrock basalt, brought to the surface of the powdery regolith by the churning of meteorite impacts.

The Taurus-Littrow basalts turned out to be vesicular, with shades of gray, buff, and pink. Analyses on Earth showed them to be very similar to the Apollo 11 basalts: they were just as old (3.7 billion years), and displayed the same high concentration of titanium oxide. This was not much of a surprise since the Apollo 17 landing site was on the edge of the Sea of Serenity, close to where it merged with Apollo 11's Sea of Tranquillity. The two sites were less than 1000 km distant, and similar lavas spilled over from one sea to the next.

Schmitt noted the large number of vesicles in the rocks he was sampling – holes that popped forth in the cooling lava as it exsolved its gas. Some boulders were layered with contrasting beds of different-sized vesicles, a stratification brought about by the cooling and flow history of the rock. The astronaut also made out two distinct generations of lava: buff, coarse-grained basalts; and bluish-gray, fine-grained basalts. The bluish variety was often found incorporated as lenticular fragments within the buff variety, as if the buff type were younger, and had engulfed fragments of the older bluish type when it erupted onto the surface.

At the close of their first outing, Cernan and Schmitt returned to the LM with a haul of 13 kg of samples. A lot of ancient basalt, but no trace yet of the fresh-looking ash they were hoping to find...

Discovery of the orange soil

The astronauts' second outing, set to last seven hours, was dedicated to an extensive survey of the valley. Four major stops were scheduled on the loop, the first one being the avalanche apron at the foot of South Massif.

The ride to the mountain front took close to an hour, and once they reached their objective, Cernan and Schmitt set out to collect samples and make observations. As expected, the avalanche boulders were breccia – 'puddings' of ancient, crustal rock fragments, mixed and welded together by the battering of giant meteorites in the early days of lunar history. Some of the fragments were rich in olivine, and pointed to very early phases of lunar magmatism (see fig. 4.5). Schmitt estimated that the boulders were recently detached from the rim of the massif, and managed to spot their source region up the layered mountain front.

Struggling to remain on schedule, Cernan and Schmitt returned to the rover and set off to the northeast, across the avalanche apron that extended from the mountain base. Stop 4 on their valley traverse was Shorty crater, a bowl-shaped depression 110 m in diameter – roughly the size of a football stadium (see fig. 3.11).

Shorty was one of the dark-halo craters that was spotted from orbit, and

FIG. 3.11. Jack Schmitt and the Lunar Rover on the edge of Shorty crater. The mound to the right of the rover is fractured basalt uplifted by the formation of the 110 m-wide impact crater, which lies out of frame to the right. It was at the foot of the mound and in the crater wall that Schmitt discovered patches of orange soil, first believed to be hydrothermally altered minerals from a recently-active volcanic vent (see fig. 3.12 for a close-up). Credit: NASA.

tentatively identified as a volcanic cinder cone. Expectations therefore ran high when the two astronauts parked their rover on the edge of the crater. Lagging behind schedule, and their oxygen supply running low, Cernan and Schmitt were allowed only 35 minutes to explore the site.

At first, the visit to the crater was disappointing: Shorty was just one more impact crater, with its characteristic ejecta blanket – chunks of bedrock basalt exposed on the slopes. But as Schmitt was circling the rim, he stopped suddenly in his tracks, staring at the ground.

'Wait a minute!', he exclaimed, 'the soil is orange!'

Schmitt lifted his gold-plated visor to make sure the orange tint was real, and not a mere illusion. The orange tint, very volcanic-looking, did not go away. Cernan hurried over. Kicking up the soil with the tip of his boot, Schmitt uncovered a streak of bright orange soil, stretching over several meters (see fig. 3.12).

'If there is anything that looks like a fumarole, this is it!', he ventured, in reference to the bright-colored deposits that line eruptive fissures on Earth.

Schmitt guessed the volcanic soil to be only a few million years old, which would have made it close to a thousand times younger than anything else discovered on the Moon until that point. If such youth were confirmed, all models of lunar and planetary volcanism would need to be revised. The orange tint also suggested hydrothermal oxidation, which was puzzling since all samples so far had turned out completely waterless.

FIG. 3.12. Discovery of volcanic orange soil by astronauts Schmitt and Cernan, on the south rim of Shorty crater (Taurus-Littrow site, Apollo 17). The astronauts dug a trench to sample the soil, visible in the foreground below and to the right of the gnomon rod (the rod is 46 cm long for scale). The bright patches encase light gray fragmental material. The fractured rock in the background is coarse-grained basalt. In the laboratory, the orange soil was shown to be made up of glass beads from lava fountaining activity, buried by later lava flows and excavated by the recent Shorty impact. Credit: NASA, photo by Harrison H. Schmitt.

In great haste, the two astronauts gathered spadefuls of orange soil, dug a trench with their rake, and drove a core through the top layers. Schmitt described the orange pocket as measuring 4 m long and 1 m wide. The core showed it to be 20 cm deep, underlain by fine-grained black soil to a depth of at least 70 cm, where the sample ran out. The edges of the pocket of soil were sharp and vertical, with the color shading off from dark orange at the center, to yellowish orange on the margins.

Pressed for time, the astronauts managed to find and describe two more streaks of orange soil at Shorty crater: one on the rim of the crater, and the other halfway down the slope, apparently excavated by a secondary impact. But Houston was calling for a wrap, and the astronauts reluctantly boarded their rover to drive back to the LM, leaving Shorty crater and its mysteries behind...

Mystery in the lab

This exciting discovery marked the highlight of the Apollo 17 mission. Ron Evans, the third crew member in lunar orbit, managed to spot Shorty's orange tint from a cruising altitude of fifteen kilometers, with the use of his telephoto lens. And when Cernan and Schmitt fired the LM's ascent stage and tore off the lunar surface to join Evans in orbit, their west-bound trajectory took them grazing at low altitude, for one last look, over Shorty crater.

Back on Earth, the orange soil delivered by the astronauts was unpacked with great anticipation and run through the gamut of preliminary analyses. The soil turned out to be composed of myriads of microscopic droplets of volcanic glass, the size of pin heads, ranging in colour from pale yellow to near black, with many peaking in the orange part of the spectrum. Their chemical make-up was basaltic, nearly undistinguishable from that of the larger chunks of lava that were collected throughout the valley. As for the orange tint, it was due to the glassy nature of the titanium-rich beads, rather than to hydrothermal or fumarolic activity.

Similar glassy spherules, green in color (high magnesium content), were collected by the Apollo 15 astronauts near Hadley Rille. Geologists attribute the colored beads on both sites to lava fountaining – pyroclastic eruptions spraying high above the lunar surface: the magma chills in the lunar vacuum to form glass droplets, which shower all around the erupting vent. This is a well-known phenomenon on Earth, where similar pyroclasts are known as *scoria*: they are usually larger than their lunar counterparts, irregular and often porous, and some are strung out by their flight through the air into stringy filaments, known as *Pele's hair* – in tribute to the Hawaiian goddess of volcanoes.

But the biggest surprise – and a disappointing one to many – lay in the age of the Apollo 17 pyroclasts: 3.7 billion years, just as old as the other rocks on the site. The soil seemed astonishingly young, not because it really was, but because it was stored at the bottom of the volcanic pile, and only recently unearthed by meteorite impacts. Sprayed in streaks over the surface, the pyroclastic beads had not yet been altered by the relentless bombardment of cosmic rays and micrometeorites. In this respect, astronaut Schmitt was correct in estimating the age of the soil to be only a few million years: analyses of the beads' surface pointed to an exposure age of 8 million years.

So ended the Apollo program, with an extraordinary haul of data to serve both volcanologists and geologists attempting to unravel the origin of the Solar System. After twenty years of data analysis and modeling, planetary geologists can now piece together a detailed history of the Moon, and of its rich volcanic activity. We review their findings in the following chapter.

Table 3.1 *A range of lunar lavas from four Apollo sites*

	Apollo 11 (average)	Apollo 12 Quartz basalt	Apollo 12 Alum basalt	Apollo 15 Olivine basalt	Apollo 15 Green glass	Apollo 17 Titanium basalt
SiO_2	40.5	46.1	46.6	44.2	45.6	37.6
Al_2O_3	10.4	10.0	12.5	8.5	7.6	8.7
FeO	18.5	20.7	18.0	22.5	19.7	21.5
MgO	7.0	8.1	6.7	11.2	16.6	8.2
CaO	11.6	10.9	11.8	9.5	8.7	10.3
Na_2O	0.4	0.3	0.7	0.2	0.1	0.4
K_2O	0,1	0.03	0.07	0.03	0.02	0.08
TiO_2	10.5	3.3	3.3	2.3	0.3	12.1
MnO	0.3	0.3	0.3	0.3	0.2	0.2
Cr_2O_3	0.3	0.5	0.4	0.6	0.4	0.4
Total	99.6	100.23	100.37	99.33	99.22	99.48

A range of lavas on the Moon: chemical compositions measured in samples returned from four sites: the Sea of Tranquillity (Apollo 11), the Ocean of Storms (Apollo 12), Hadley Rille (Apollo 15) and Taurus-Littrow (Apollo 17). Percentage in mass (oxides). Error margins (not included) occasionally account for totals over 100%.

Lunar basalts are characterized by a relatively low silica content, and high ratios of iron and magnesium oxides. Inversely, they are poor in volatile elements such as sodium and potassium.

The Apollo 11 and 17 basalts (first and last column) both display a high content of titanium oxide (they come from adjacent maria). The Apollo 12 basalt represented here is an aluminous variety. The Apollo 15 basalt is of the olivine variety, collected near the rim of the Hadley Rille lava channel. The Apollo 15 green glasses are pyroclastic beads from fire fountaining activity: this glass appears to represent the most primitive (least differentiated) igneous product sampled so far on the Moon.

Modified from Stuart R. Taylor, *Lunar Science: A Post-Apollo View*, Pergamon, 1975.

4 Lunar volcanism

Treasures of Apollo

From the first moonfall of July 1969 to the farewell mission of December 1972, six Apollo crews roamed the lunar surface, hiking and driving over 100 km across basaltic plains, up and down crater slopes, and along fault scarps and mountain fronts. The twelve astronauts set up one preliminary, and five operational science stations; took thousands of 35 mm and 70 mm photos, and dozens of hours of video footage; and collected nearly half a ton of lunar samples (381.7 kg).

Two thirds of the sample haul consisted of rocks and rock fragments ranging in size from a few grams to a record nine kilograms for the largest specimen – the cantaloup-sized 'Great Scott' basalt collected by astronaut David Scott. The balance of the samples consisted of spadefuls and rakefuls of lunar soil – myriads of sand-sized mineral and glassy fragments. This soil, named the *regolith*, results from the constant pounding and churning of the lunar surface by meteorite impacts.

The regolith is particularly interesting in that it represents a natural 'potpourri' of lunar mineralogy, containing both fragments of the local subsurface, and foreign rock debris balistically emplaced from impact sites hundreds of kilometers away. The astronauts collected 90 kg of regolith on six sites, including several cores sampling the soil layers in cross-section. One of the Apollo 15 cores penetrated 2.4 m

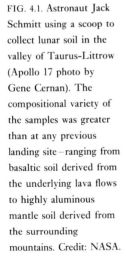

FIG. 4.1. Astronaut Jack Schmitt using a scoop to collect lunar soil in the valley of Taurus-Littrow (Apollo 17 photo by Gene Cernan). The compositional variety of the samples was greater than at any previous landing site – ranging from basaltic soil derived from the underlying lava flows to highly aluminous mantle soil derived from the surrounding mountains. Credit: NASA.

into the ground and sampled fifty distinct layers (1 to 10 cm thick), each representing a discrete blanket of ejecta from an impact near or far.

Analyzing the regolith tells us something about the nature and dynamics of impact processes on the Moon, while at the same time broadening the scope of our mineral sampling. This is akin to sampling pebbles in a terrestrial riverbed: the range and variety of samples is enhanced, because it is assembled from a variety of geological provinces upstream. On the Moon, impacts play an equivalent, double role of erosion and distribution.

In effect, only two great forces shape the Moon's geology and landscape: magmatism, which initially melted and stratified the outer layers of our satellite, and fueled several pulses of volcanic activity at the surface; and meteorite bombardment, which shattered and uplifted crustal blocks, and redistributed mineral material as ballistic ejecta and base surges over the landscape. Both these forces – volcanic and meteoritic – were most active during the first billion years of lunar history, and declined sharply thereafter.

A capsule history of the Moon

Based on our current knowledge, the history of the Moon was played out in four major acts, beginning with a most extraordinary prelude.

The prelude is the creation of the Moon itself. Geochemical constraints have led an increasing number of scientists to believe that the Moon was born out of a gigantic collision, in the early days of the Solar System. As the planets were accreting around the Sun, 4.6 billion years ago, a planetary body roughly the size of Mars collided with the young Earth, vaporizing into a stream of matter that condensed into a ring system around our planet. The ring debris swiftly coalesced to form a single body – the Moon – in a runaway accretion that brought its growing envelope to very high temperatures (see fig. 4.2).

From then on, the creation of the Moon – or should we say its 'reincarnation' – proceeded along the normal lines of planetary evolution. The outer layers partially melted as a result of the elevated accretion temperatures, and the hot mineral mush underwent large-scale differentiation, leading to the emplacement of a primeval lunar crust. Melting was so pervasive that this first era of lunar history is known as the era of the 'magma ocean'.

The turbulent childhood of the Moon graded into a second era, where the crust was now solid, but continually disturbed by unabated meteorite bombardment, and presumably a good deal of igneous activity (much evidence of lava flows from this period was shattered and buried by later impacts and ejecta). This era of brisk 'sweeping up' of cosmic debris shaped the cratered highlands of the Moon, and the larger impacts blasted prominent basins into the lunar crust, surrounded by rings of faulted-up mountain blocks. This second act of lunar history lasted from approximately 4.2 to 3.8 billion years ago, ending with the excavation of the last two giant basins – Imbrium and Orientale.

After this heavy bombardment phase, the Solar System quieted down. The

FIG. 4.2. A large-scale collision between the Earth and a smaller planet might have been responsible for the formation of the Moon in the early days of the Solar System. This would explain why the Moon, made predominantly of material blasted out of the Earth's upper mantle, is globally iron-poor (the Earth's core was not involved), while both bodies were otherwise endowed with striking isotopic similarities. Credit: Painting by William K. Hartmann.

meteorite flux declined, and the Moon was affected more by its ongoing internal processes than by outside intervention. The build-up of radioactive heat at depth led to melting episodes, and batches of magma made it to the surface through the crustal cracks of the large basins. The basins were repeatedly flooded by lava flows, separated by intervals of inactivity. This third era of lunar history – the epoch of mare flooding – covered the interval 3.8 to 3 billion years ago, and probably ran into the following aeon: late mare flows in the Imbrium basin are speculated to be as young as 2 billion years.

By then, the Moon had expended the bulk of its radioactive heat, and volcanic activity dwindled to a trickle. The last 2 billion years were disturbed only by small and rare magma pulses and degassing events – constrained to a few areas in Mare

Imbrium and the Ocean of Storms – and by the occasional blast of a late-falling asteroid.

The Apollo missions brought samples back to Earth covering all four periods of lunar history: Apollo 14 and 16 secured fragments of the Moon's primeval crust, and breccia from the heavy bombardment phase; Apollo 11 and 12 collected basalt from the volcanic era of mare flooding; and Apollo 15 and 17 scored on all counts, bringing back samples of the early crust, volcanic mare basalts, and glassy breccia from recent impact craters.

Sampling strategies

Our knowledge of lunar history and volcanism relies principally on the six near side sites visited by the Apollo astronauts. Despite the unavoidable mission constraints – due to fuel range, and safety considerations – the site selection was remarkably effective in allowing as broad a coverage as possible of the main stratigraphic units on the Moon, different periods of its history, and major questions and issues of lunar geology.

One limitation was that all sites had to be chosen on the Moon's near side, in order to be in continuous radio contact with Earth. The handicap turned out to be minor, because most of the interesting geology occured precisely on the Moon's near side.

Apollo rock and soil samples total 382 kg, to which we must add three small drill cores collected and returned automatically by the Soviet probes Luna 16 (1970), Luna 20 (1972), and Luna 24 (1976). The Soviet strategy was both clever and economic. It also came close to upsetting the Americans in the Moon race, had Luna 15 not failed in the first sample-and-return attempt of July 1969. This first probe of the series crashed in the Ocean of Storms, the very day Armstrong and Aldrin descended to the lunar surface. Had it performed as planned, Luna 15 would have returned a Moon rock back to Earth one day before the Americans, and at one hundredth the cost!

The automatic operation – successfully demonstrated by Luna 16 one year later – consisted in soft landing a two-stage probe: the landing stage drilled the soil beneath it, and transferred the core to the upper stage, which then rocketed back to Earth. Stored in a reentry capsule, the sample was delivered safely by parachute to the recovery team, somewhere in the steppes of Kazakhstan.

The samples brought back by the Soviet probes were certainly modest in weight (100 grams, 30 grams, and 170 grams respectively), but they broadened our sampling range by adding three new sites to our growing list, all on the eastern limb of the Moon: the Sea of Fecundity (Luna 16), the Sea of Crises (Luna 24), and the Apollonius highlands separating the two basins (Luna 20).

Classification of the lunar lava

Samples are thus available from nine different sites on the Moon's near side (see fig. 3.3). On the mare sites (six sites out of nine), basaltic lavas are the dominant rock family, and from west to east, show subtle changes in age and composition.

To the west we find 'young' basalts, sampled by Apollo 12 in the Ocean of Storms,

FIG. 4.3. A vesicular olivine basalt, collected at Hadley Rille by the Apollo 15 astronauts. The sample measures 13 cm and weighs 924 g. The abundant vesicles point to degassing of volatiles from the magma, probably carbon monoxide. Lavas from Hadley Rille were dated at 3.3 billion years. Credit: NASA/Johnson Space Center, courtesy of Dr James L. Gooding.

dated at 3.2 billion years. These are the youngest lavas collected so far on the Moon (although we know some mare regions are much younger), and they can be split into three categories: basalts with crystals of feldspar and pyroxene; a second variety containing olivine as well; and a third type boasting high titanium content. Moreover, the fine-grained regolith at the site was found to contain mineral shards and glass of a much older type of magma, pulverized and mixed into the soil by aeons of meteorite pounding.

Three hundred kilometers to the east, the Apollo 14 astronauts found sizeable chunks of these older lavas preserved in the breccia rock of the Fra Mauro Formation, a hummocky ejecta blanket dating back to the heavy bombardment era. These early basalts are aluminous in nature, and are dated at 3.9 billion years.

Five hundred kilometers farther to the east (close to the zero meridian), and up in the northern latitudes, at the intersection of the Imbrium and Serenitatis basin rings, the Apollo 15 astronauts sampled more lavas typical of mare volcanism – the Hadley Rille basalts, on the edge of Mare Imbrium. The astronauts brought back two varieties: a pyroxene basalt making up the bulk of the flows, and an olivine-rich specimen from the uppermost layer (see fig. 4.3). The Hadley site lavas are slightly older than those sampled to the west: 3.3 to 3.4 billion years. An older, aluminous type of basalt turns up on this site as well, buried on the floor of the basin and scantily exposed along the rim (the Apennine Bench Formation).

There is a significant change in the age and chemistry of samples when one crosses over to the eastern quadrant of the Moon's near side. The basalts brought back from the Sea of Tranquillity (Apollo 11) are characterized by a high titanium content – up to 12% in weight. They are also significantly older than the lavas from the western quadrant: 3.7 to 3.8 billion years, a difference of some 500 million years.

These high-titanium basalts are divided into subcategories, based on their

FIG. 4.4. A thin section of lunar lava (Apollo 15 'Great Scott' olivine basalt), viewed under the polarizing microscope. The width of field is 4 mm. The mosaic of crystals includes large phenocrysts of clinopyroxene (center) laths of plagioclase feldspar (black and white striped parallelograms); small pearly grains of olivine; and uncrystallized glass, ilmenite and other metal oxides (black specks). Credit: NASA/Johnson Space Center, courtesy of Dr James L. Gooding.

potassium content (low, intermediate, or high), with a fourth, slightly younger class (3.6 by) containing less titanium than the rest. This is a most diverse set of samples, when one considers that Armstrong and Aldrin gathered their collection in little more than 20 minutes, and within 50 m of the Lunar Module!

Apollo 17, the last lunar mission, retrieved lavas of similar age and composition one thousand kilometers north of Tranquillity Base, along the edge of the Sea of Serenity: there, the local basalts were also aged 3.7 billion years, and contained much titanium. Cernan and Schmitt also brought back very ancient igneous rocks and lava clasts from the surrounding mountain fronts.

Finally, the Russian Luna probes sampled the easternmost lava plains of the Sea of Fecundity and the Sea of Crises, rocketing back fragments of aluminous basalt that turned out to be intermediate in age (3.4 by).

The chemistry of the lunar lava

What conclusions can we reach from the rock collection at hand?

First of all, that most mare lavas are basaltic and dry in nature. They all display ferromagnesian and calcic chemistry, with moderate aluminum and silicon. Minerally speaking, this translates into a mosaic of feldspar and pyroxene crystals, with a sprinkle of iron-titanium oxides, and occasional olivine or quartz (see fig. 4.4).

On the other hand, there are virtually no lavas on the Moon more siliceous or alkaline than basalt – no andesites, dacites, rhyolites and the like (except for a few clasts), whereas such differentiated lavas are relatively common on Earth, where they represent close to 10% of all lava exposed. This absence of differentiated lavas on the Moon is essentially due to the quick ascent and extrusion of the magma, leaving little time for differentiation to take place. Also, water plays a key role in differentiation – at least in many siliceous systems on Earth – and the absence of water in lunar magma certainly does nothing to help.

Although they are restricted to only the basalt family, lunar lavas embrace a wide range of subcategories. For one, petrologists divide lunar basalts into a range of chemical groups, on the basis of their titanium content. Titanium is a natural choice as a marker because its concentration varies dramatically across the basalt collection, from high-titanium basalts ($>6\%$ TiO_2); to medium-titanium basalts (6 to 2% TiO_2); low-titanium basalts ($<2\%$ TiO_2); and a very-low-titanium class (VLT basalts). As a whole, titanium ranks seventh in the list of lunar constituents (behind oxygen, silicon, aluminum, iron, magnesium, and calcium).

Potassium is another element that displays high variability across the board, and is used to further subdivide the lavas. Finally, extra classes of aluminous basalts round off the list.

Other classification schemes are used for lunar basalts, based on minerals rather than chemistry (feldspathic basalts, cristobalite basalts, etc.); or texture considerations (size of crystals, presence and size of vesicles and vughs).

Besides their range in chemistry, mineralogy, and texture, lunar lavas also vary significantly in age: the sampled mare basalts cover 600 million years of lunar volcanism, from 3.8 to 3.2 billion years ago. Extending this range, the clasts of aluminous basalts found in the highland breccias testify to older pulses of volcanism, dating back to 3.85 and 3.9 billion years (Apollo 15 and Apollo 14 clasts), and perhaps even 4.2 billion years (Apollo 16 clasts). On the other end of the scale, patches of basalts as young as 2 billion years have been inferred in a few mare areas, on the basis of crater-count dating.

Chemical and age variety should not eclipse the fact that mare basalts share a number of common characteristics, which give them a unique lunar 'flavor', and set them apart from their terrestrial counterparts.

First and foremost, lunar rocks and minerals totally lack water. This is in sharp contrast to the Earth, where water enters the framework of numerous minerals – mostly as hydroxyl ions OH^-, which act as binding agents in crystal lattices: on Earth, close to 90% of the 2000 described mineral types contain water, most notably the mica, amphibole, and clay families.

Deprived of water, lunar mineralogy is restricted to a much narrower range of crystalline arrangements: only one hundred mineral species have been identified so far, a far cry from the Earth's two thousand! Heading the list are the fundamental – and anhydrous – building blocks of igneous rocks: pyroxenes, feldspars, and olivines; followed by the metal oxides magnetite and ilmenite; and by quartz, and zircons.

The list is short, but it does include a few exotic species not found on Earth, and which owe their existence to the Moon's reducing chemistry. One is a new form of titanium oxide with iron and magnesium – baptised *armalcolite* in tribute to Apollo 11 astronauts Armstrong, Aldrin, and Collins. There is also a complex silicate of iron and titanium named *tranquillityite* in reference to the Sea of Tranquillity; and a reduced silicate of iron and calcium christened *pyroxferroite*. Since their discovery in lunar samples, some of these minerals have also been found on Earth, in rocks formed in very special environments.

Besides their totally anhydrous nature, and their reduced rather than oxidized state, lunar rocks are further distinguished by their extreme dearth of volatile elements: not only do they lack water, but they are also very poor in sodium, potassium, and other elements with a low melting point. Inversely, lunar rocks are particularly rich in refractory elements – those with a high melting point – such as titanium and chromium.

This contrasting chemical make-up between the Earth and Moon could be the direct result of the postulated planetary collision that gave birth to our satellite in the early days of the Solar System: in the jet of vaporized matter that ensued, volatile elements were lost to space, and the matter that condensed to form the Moon was thereby enriched in the more refractory elements. Volatile depletion also occured in the magma ocean era, as large impacts bashed, splashed, and overturned the molten 'batter'.

The europium mystery

A lot can be learned about the evolution and igneous history of the Moon, if one looks at some of the more unusual atoms that are partitioned between the various lunar minerals.

A particular class of atoms – the incompatible elements – are especially interesting in that their concentrations and ratios in a lava reflect the various episodes of melting and crystallization that led to its formation. Because their rather odd sizes and electrical charges prevent them from fitting snugly into crystal lattices, incompatible elements are barred from entering the first crystals of a cooling magma, and end up concentrated in the last dregs of melt – in the last erupted or crystallized rock. If such differentiated rocks later remelt, incompatible elements will again be preferentially squeezed out over other elements, and further concentrated in the next generation magma.

Hence, it is possible to guess from a rock's concentration of incompatible elements if it is a one-time, pristine lava from a unique episode of melting at depth (normal content of incompatibles), or if it is the end result of a succession of magmatic episodes through time (the concentration of incompatibles rising at each step).

One particular set of incompatible elements is most instructive in this respect: the Rare Earth Elements – REE for short. They comprise the fourteen elements of the periodic table between atomic number 58 (lawrencium) and 71 (lutetium). In lunar basalts, the rare earths collectively add up to only a few thousandths of one per cent – a small proportion indeed, but ten to a hundred times larger than the value assumed for the deep lunar mantle. This ten-fold to one-hundred-fold enrichment indicates that lunar lavas are the end result of a number of melting and crystallization episodes since the creation of the Moon, especially those lavas with the largest enrichment factors.

The Apollo 15 basalts are the least enriched of the lunar samples (their rare earths are concentrated by a factor of only 10 to 20), which means that the Hadley Rille source magma was derived quite directly from undisturbed mantle at depth, and most closely reflects the mantle's composition.

Next come the Apollo 17 basalts, which are slightly more enriched in REEs (30 times the norm): their magma is thought to have undergone a somewhat more complex evolutionary history.

Finally, potassium basalts and aluminous basalts are the most enriched of all (50 to 100 times the REE norm): their source magmas were derived from regions in the lunar interior which themselves had undergone considerable differentiation prior to the eruptions.

The family of rare earth elements also holds some remarkable information about the evolution of the Moon as a whole. If one looks at the family's atomic species individually, there is a puzzling dearth of the element europium with respect to the other REE members in the lavas. Apparently, some event early in lunar history 'absconded' the europium from the source mantle regions, and the lavas that were later generated in those regions had very little europium left to carry up to the surface.

Where was the missing europium hiding? What mysterious process had affected the lunar interior early in its history?

The mystery of the missing europium was solved when the astronauts brought back samples of primeval crust from the lunar highlands. There was the europium, locked up inside the feldspar crystals of these light-colored rocks. It so happened that in the reducing environment of the Moon, europium ions were of just the right caliber and electric charge to fit into the crystal lattice of plagioclase feldspars, and nearly all of the europium did just that, ending up in the highland crust. That the migration process was so thorough meant that a phenomenal episode of melting must have affected the early Moon down to a depth of several hundred kilometers – a global 'magma ocean'.

As this grandiose 'magma ocean' cooled, feldspar minerals crystallized – pumping europium atoms out of the bath – and rose by buoyancy to the top of the incandescent ocean, where they collected to form a feldspar-rich, europium-rich anorthosite crust. When the hotter, lower layers of the magma ocean crystallized in turn to form the mantle below, there was virtually no europium left for their own crystals to host. As a result, when the mantle underwent new cycles of melting millions of years later, a dearth of europium characterized the erupted lavas.

This is how an obscure little atom of the rare earths – present only in trace amounts in the lunar samples – yielded one of the most important discoveries in planetary geology: that the outer envelope of the Moon started off thoroughly molten as a magma ocean hundreds of kilometers deep, and that it cooled and decanted over time to form the layered mantle and crust.

KREEP: the evasive lavas

This model can be refined in the light of some unusual chemistry discovered in the mare soils. In the first soil samples brought back from the Sea of Tranquillity (Apollo 11) and the Ocean of Storms (Apollo 12), it appeared that some exotic mineral grains in the regolith contained high levels of potassium (K), the rare earth elements (REE), and phosphorus (P). Bunching the atomic symbols together, geochemists came up with the anagram KREEP to describe this odd soil component.

KREEP soil fragments had to come from yet undiscovered rock types, somewhere in the vicinity of the landing sites – rocks that had been pulverized by meteorite impacts and mixed as fragments into the regolith.

KREEP particles told us two things: for one, they were enriched in incompatible elements, so that their source was a highly differentiated region of the Moon. The second thing was the age of this chemical dust – easily measurable by radiochronology – which consistently registered 4.35 billion years, thereby representing the creation of an original, widespread and differentiated magma at that time, only 200 million years into the history of the Moon. Was there an ancient type of lava yet to be discovered – a KREEP basalt – and if so, where was it hiding? And where did it come from?

The Apollo 14 mission, landing in the hummocky hills of Fra Mauro, helped to lift the veil. Fra Mauro is an area of rolling hills on the eastern edge of the Ocean of Storms, and its origin was the object of much debate, prior to the mission. Those who believed the features to be volcanic were disappointed, for astronauts Shepard and Mitchell collected mostly crustal, broken-up breccia from the asteroid or comet blast that excavated the Imbrium basin to the north. But their sample bags also turned up some of the first clasts of lava with KREEP characteristics – KREEP basalts.

These were aluminous, potassium-rich basalts, with high KREEP concentrations. Their crystallization age registered around 3.9 billion years, not the older 4.35 by figure obtained from the KREEP component individually. Moreover, another set of KREEP basalts returned from the Apennine foothills by the Apollo 15 astronauts – the Apennine Bench Formation – also came out under the 4 by mark, registering 3.85 by.

What this all meant would fall in place when lunar geologists picked up the pieces, and started assembling a global picture of the Moon's igneous history.

Act One: the magma ocean

The Moon, as we saw in our capsule history, is believed to have accreted from a ring of debris strewn around the Earth by a head-on planetary collision of our planet with another medium-sized body in the formative years of the Solar System. The collision debris, scattered in a ring around the Earth, rapidly and violently accreted to form the Moon, and the growing lunar surface quickly reached temperatures in excess of 1400 °C (1700 K), heated by impacts from above and radioactive decay from beneath.

The europium anomaly tells us that the melting of the Moon's outer shell was thorough, with a bubbling ocean of magma hundreds of kilometers deep – 500 km

FIG. 4.5. This lunar sample, which bears number 72415 in the Houston collection, is the oldest rock ever found on a planet, dated at 4.5 billion years. It is a dunite – an igneous cumulate made of 90% olivine – and was collected by the astronauts of Apollo 17 from a breccia boulder at the foot of South Massif. The dunite is assumed to have crystallized out of a deep magma ocean in the early days of lunar history. Credit: NASA/Johnson Space Center, courtesy of Dr James L. Gooding.

being an average estimate. This gigantic planetary cauldron took close to 200 million years to cool and solidify completely (4.55 to 4.35 by ago), the atoms linking into crystals and migrating through the broth to stratify in layers, according to differences in density and to temperature and pressure gradients.

In places, an outer veneer, rich in magnesium, congealed along the bottom and roof of the global magma cauldron, and in smaller magma pockets. A magnificent sample of this very early igneous rock was returned by Cernan and Schmitt from the Taurus-Littrow mountain front: this pale dunite – a greenish rock almost exclusively composed of magnesium-rich olivine – registers 4.5 by, truly the oldest rock ever found on a planet (see fig. 4.5).

The bulk of the magma broth, meanwhile, began spawning light crystals of plagioclase, which rose to form the lunar crust; and heavy 'slag' crystals of olivine and pyroxene, which settled to form the bottom layers, as the magma ocean slowly cooled and solidified.

This deep, igneous cauldron was far from a simple system, however. Samples brought back from the crustal highlands are not chemically uniform, and include both anorthosites of almost pure plagioclase composition (refered to as the ferroan anorthosite suite, for they do contain a hint of iron silicates) and a second anorthosite family which contains magnesium-rich olivines and pyroxenes, known as the Mg-suite. This second family reflects more localised magmatic processes, also occuring at the time of the magma ocean.

Indeed, the magma ocean gave birth to a complex and unstable system. Its

FIG. 4.6. Schematic cross-section of the evolution of the Moon as a function of time. In the beginning (4.5 by ago), the Moon was molten to a depth of several hundred kilometers (black). This magma ocean progressively solidified, while the thickening crust was shattered by impacts (megaregolith). By 4.3 by ago, only a thin lens of residual melt was left at a depth of a few hundred kilometers – the source of KREEP basalt. By 4.0 by ago, the residual magma ocean had completely solidified, while radioactive decay triggered melting deeper in the mantle – the source of the mare lava flows. After 2 by ago, this partially-molten layer was constrained to such depths that lava was no longer able to reach the surface. Credit: From William K. Hartmann, 1983, *Moons and Planets*, Wadsworth Publishing.

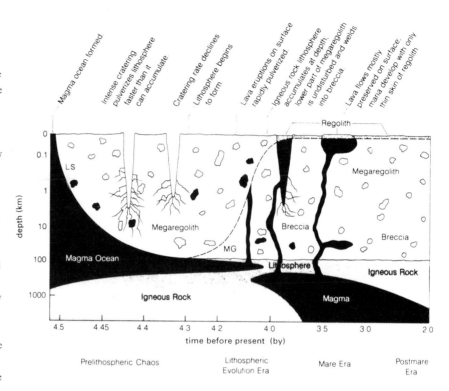

crystallized upper layers ended up overlying hotter and more buoyant mantle at depth, so that the pile might have occasionally overturned, sending pockets of mantle material upwards, and sinking slabs of dense crust downwards (see fig. 4.6).

The slow solidification and differentiation of the magma ocean – disturbed by local or global overturns – lasted close to 200 million years, at the end of which there were only thin lenses of residual liquid left between the olivine-rich dregs and the plagioclase-rich crust. These last lenses of decanted magma, thirty to forty kilometers beneath the surface, had become enriched in aluminum, potassium, and incompatible elements, and resulted in the peculiar KREEP chemistry which later showed up in the Apollo samples.

According to the magma ocean model, the KREEP layer was isolated 4.35 by ago: the KREEP particles of that age found strewn across the lunar surface are pulverized fragments of the layer, dug up to the surface by meteorite impacts. As for the younger 3.9 and 3.85 by KREEP basalts, sampled by the crews of Apollo 14 and 15, they are local remelts of the KREEP source region which rose to the surface some 500 million years after the formation of the layer at depth.

The model of a stratified magma ocean in the hot, early history of a planet is a tantalizing concept. Geologists are curious to find out if the Earth also began its history with a frothing mineral bath hundreds of kilometers deep. There is reason to suspect so, since the Earth apparently experienced a planetary genesis similar to the Moon's, notably the energetic planetary collision which gave birth to our satellite. No doubt such a collision would have melted a great proportion of the Earth, perhaps as much as 50%, creating magma oceans hundreds to thousands of kilometers in depth. Unfortunately, the Earth's oldest rocks only reach back 3.8 to 3.9 billion years – a few isolated minerals go back a couple of hundred million years further – so that there is virtually no record of the first 500 million years of the Earth's history.

If an early magma ocean did in fact exist on Earth, it certainly did not leave many clues for us to study, although there is some chemical evidence that a plagioclase-rich crust – expected of a magma ocean – might have existed on the early Earth. Alternatively, it is argued that a terrestrial magma ocean might not have left a tell-tale plagioclase crust at all. Indeed, the magma ocean is expected to have been hundreds of kilometers deeper on Earth than on the Moon, and the resulting higher pressures would have steered crystal growth in a radically different direction than in the lunar case. Rather than aluminum and calcium entering plagioclase feldspar and rising to the top of the crust, those same elements could have locked into dense crystals of garnet, and sunk to the bottom. Not enough aluminum and calcium would then have been left to build a plagioclase 'scum' that would survive through the aeons, and betray the past existence of a magma ocean on Earth.

Had it not been for the Moon and its magnificent feldspar-packed highlands, we might never have guessed what churning inferno ruled the early days of the terrestrial planets. . .

Early volcanism and impact melts

The lunar magma ocean developed and stratified under a hellish rain of meteorite blasts – high energy impacts stirring the cauldron and adding new mineral stock to it. With time, a primeval crust did form over its cooling surface – as we just discussed – although impacts kept pounding and remelting that 'lid' on a regional scale. Molten pockets breached through to the surface, and volcanism was certainly widespread in this turbulent era of lunar history.

There is good reason to believe that the KREEP layer, at the top of the mantle, provided most of the early magma pockets: not only were the KREEP lenses last to crystallize from the magma ocean, but they most certainly flipped back into a partially molten state soon thereafter, because of the quantity of radioactive elements – uranium, thorium, and potassium – which they contained. Evidence of early KREEP lavas were found by the Apollo astronauts as clasts on the Descartes site (dated at 4.2 by); in the breccias of Fra Mauro (3.9 by); and as whole rocks at the base of the Apennine mountains (3.85 by). Magmatism of KREEP chemistry was therefore prevalent from 4.2 to 3.85 by ago, and took advantage of major impacts to seep up through the faulted crust, and flood basin floors.

Indeed, the meteorite flux at the surface was showing no signs of abating. On the contrary, it apparently underwent a surge as the Moon 'cleaned up' the larger, asteroid-sized objects still crossing its orbit. This collision surge is known as the Heavy Bombardment Period, during which large objects blasted deep basins, with shock rings and foundering terraces. From crater counting and sample dating, it appears that the Nectaris and Humorum basins marked the beginning of this

bombardment surge 3.92 by ago, followed by over a dozen large events, including Crisium and Serenitatis (3.87 by), and ending with Imbrium (3.85 by) and Orientale (3.80 by) – the last two giant impacts.

In the midst of this bombardment, earlier lava flows and crustal blocks were broken up and their debris tossed about the lunar surface, and to complicate the picture, sheets of impact melt were formed by the blasts, flowing in surge-like fashion, and pooling in newly formed basins and depressions. The hummocky, brown patches in northeastern Orientale (the Maunder Formation) are believed to be such impact melts, well preserved because of their relative youth: they are believed to be one kilometer thick. Melt sheets of older basins are more difficult to detect: they have been covered by younger ejecta and mare lavas, or intricately mixed in breccia 'puddings', as observed in the boulders of the Taurus-Littrow site (see fig. 4.7).

Impact and volcanic processes are often difficult to tell apart: the emplacement of impact melt closely mimics volcanic processes, and deposits that were once thought to be volcanic pyroclasts (such as the smooth Cayley plains in the lunar highlands) are now thought to be impact ejecta, fluidly emplaced in a cloud of hot gas, as are ignimbrite sheets on Earth.

Impact melts can also undergo differentiation, and resemble volcanic suites. We have an interesting example on Earth with the Sudbury complex in Ontario, long mistaken for a volcanic structure. The eroded, 1.8 by old crater displays a ring of melt rock averaging 50 km in width and 2500 m in thickness. The large melt pool took on the order of 100 000 years to cool, time enough to undergo fractional crystallization: the complex is layered, with mafic norites at the bottom and plagioclase-rich granophyres at the top.

In the lab, however, impact melts stand apart from their volcanic, deep-derived cousins. First, they reflect the composition of the local target rocks which they melted. An impact in the anorthosite highlands of the Moon will yield a melt pool rich in aluminum and calcium; a blast in mare lava flows will stir up a melt looking like basalt, and full of iron-rich glass. And secondly, chemical, isotopic, and textural evidence give away the meteoritic origin of such 'instant melts'.

Remote-sensing from orbit has revived the search for impact melts, singularly with the Galileo flybys of 1990 and 1992, and the Clementine mission of 1994. Iron-bearing glass shows up red in multispectral imagery, and is prominent along the rims of Plato, Tycho, Copernicus, Jackson, Kepler, Archimedes, Langrenus, and many other craters. The concentration of glass along the rims is believed to reflect the mechanism of crater formation, by which impact melt is thrust up along ring faults.

During the Heavy Bombardment Period, impacts might also have directly triggered the upwelling of molten KREEP material, which underpinned the thin crust. One example is the volcanic KREEP basalt of the Apennine Bench Formation, collected by the Apollo 15 astronauts on the rim of the Imbrium basin and dated at 3.85 by, which is exactly the age of the blast that dug the basin. Apparently molten KREEP at the base of the crust was stripped by the impact, surging through the fractures into the newly created basin.

Mare volcanism

As the Heavy Bombardment Period came to a close, 3.8 billion years ago, the Moon shifted into a phase of dominant volcanism: impacts became smaller and scarcer, and deep-seated lavas were left undisputed, rising massively to the surface of basins to create the maria.

The maria are the smooth, dark plains that cover nearly a third of the Moon's near side – close to 6 million square kilometers – and are found in a few patches on the far side. There was a time, before Apollo, when they were thought to be impact melt, created by the blasts that formed the giant basins. Speculations of impact origin were dispelled when most samples returned to Earth proved to be deep-source basalts, hundreds of millions of years younger than their enclosing basins. Mare basalts were therefore not the immediate result of basin excavation, but later, independent pulses of magma which took advantage of the low lithospheric burden, and the fractured crust which established convenient pathways to the surface.

Indeed, magmatism was proceeding at depth, contemporaneous with the battering going on at the surface. In the lower crust, no sooner had the last dregs of magma ocean crystallized – shutting down the early pulse of KREEP volcanism – than the build-up of radioactive heat at greater depths sparked new cycles of melting, this time well within the mantle. It is interesting to note that the impact era that shattered the crust might have helped the energy build-up, by establishing a more insulating cover over the mantle (fissured and brecciated rock is less conductive) and forcing the temperature rise at depth.

The first mare lavas are almost all covered up by later flows and basin ejecta – as are impact melt sheets – but in some places the buried lava soil is churned up to the surface by later impacts, flashing a distinct, mafic signature. These early mare basalts, betrayed by their churned-up soil, are known as *cryptomaria*. The recently discovered South Pole–Aitken basin, which dominates the southern far side of the Moon, is believed to contain such a cryptomare. Other early basalts, broken-up and mixed with ejecta, have been detected by remote-sensing in the Schiller–Shickard area (southwestern limb), the Lomonosov–Fleming basin, and the polar plains north of Mare Frigoris.

From these early cryptomaria to the youngest sampled basalts, mare volcanism spans close to a billion years, from over 4 by to under 3.2 by ago. Moreover, if we turn to crater counts and other stratigraphic clues, we can see that mare volcanism extended past the 3 by mark, and that some mare patches are not much older than 2 by, namely near the Flamsteed Ring of the Ocean of Storms. There are even a few rare flows that embay such sharp-looking craters that their age is estimated to be close to 1 by – which would make them very young flows indeed.

We turn to remote-sensing techniques to speculate on the age and nature of these unsampled lava fields. From the wider perspective of lunar orbit, mare volcanism is seen to be quite varied, in keeping with its long eruptive history. For example, aluminous basalts are much more common than our returned samples could lead us to believe. The orbital surveys show them to be widespread in the Moon's eastern quadrant (the right side, as viewed from Earth), confirming the aluminous sample returns from the Soviet probes Luna 16 and 24.

In contrast, the western quadrant seems to be richer in radioactive uranium and thorium, as recorded by gamma spectrometry over Mare Imbrium; Archimedes crater; Aristarchus crater and plateau; and the Fra Mauro region of Apollo 14. These western basalts apparently disposed of more radioactive fuel than their eastern counterparts, which might be the reason for longer-lasting volcanism.

Orbital surveys of the mare flows in visible light also bring out subtle color variations, linked to their titanium content: low-titanium basalts (less than 2% TiO_2) display a reddish hue; whereas medium-titanium basalts (2 to 6%) peak in the orange; and high-titanium basalts (8 to 14%) peak in the blue.

A good example of mare complexity is provided by the Sea of Serenity, which displays all three types: titanium-rich basalts in the south – this class was sampled by Apollo 17 at Taurus-Littrow; reddish, titanium-poor basalts in the basin center; and two orange-tint crescents both east and west, reflecting intermediate-titanium values.

The Serenity eruptions were also staggered in time, as is apparent from crater counts on the surface flows: the first lavas to erupt (the oldest) were the 'blue' titanium-rich basalts in the south, dated at 3.7 by (Apollo 17 samples). Next were the orange, average-titanium basalts, which outcrop as crescents on the rim of the basin. Last were the titanium-poor basalts in the center, which spanned a relatively long period of emplacement, from 3.5 to 2.5 by (see fig. 4.8).

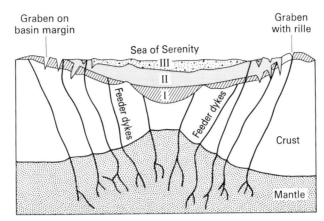

FIG. 4.8. Schematic diagram of mare flooding in the Serenity basin. The magma from the mantle rose through radial and concentric fractures created by the giant impact. The older flows, rich in titanium (phase I) are still visible on the basin margins. Later flows, poorer in titanium (phases II and III), were emplaced above them, in the more central regions of the basin. Credit: Adapted from the *Lunar Sourcebook*, Cambridge University Press.

This chemical evolution, from early titanium-rich to late titanium-poor basalts indicates that, at least in the Serenity basin, the source of mare basalts got deeper with time, migrating from titanium-rich superficial layers – where titanium minerals are stable – to lower titanium-depleted sources. This is not unexpected since the thermal evolution of the Moon commanded a progressive solidification of the upper mantle, and a migration of the partially molten zone to ever-increasing depths.

There are exceptions to this rule, however. In the Imbrium basin, lavas do start out titanium-rich (southwest of Archimedes crater), and are followed by titanium-poor batches (Apollo 15 Hadley lavas, dated at 3.4 by), but then the trend reverses and the youngest lavas of the southwest surge back to high-titanium values. Aluminous basalts also depart from a simple time sequence: high-aluminum varieties are found both in the oldest flows (the 4.2 by old Apollo 16 clasts), and in the youngest ones (the 3.4 by old Luna 24 basalts).

Moon-wide trends

Mare volcanism on the Moon was therefore affected by complex regional trends. Certainly the underlying mantle, source of the basalts, varies in composition from place to place, because the mixing of the lunar batch at the time of the magma ocean was far from complete, and preserved a great deal of lateral and vertical heterogeneity in the areas of magma genesis. This variability was certainly reinforced by mixing effects – melting and assimilation of layers crossed by the magma on its way to the surface. The relatively hot and shallow KREEP layer, in particular, certainly contaminated many a batch originating at lower levels.

Beyond its regional variability, lunar volcanism does display two major, planet-wide trends. Firstly, most mare outcrops occur on the Moon's near side, facing Earth. The far side is dominated by crustal highlands, and although it does bear the scars of major impact basins, shows few apparent flow fields – limited to Mare Australe, the basin of Moscow, the Sea of Ingenuity, Mare Orientale, and Tsiolkovsky crater.

This marked asymmetry between near side and far side might have something to do with the Earth's gravitational pull. Tidal forces acted as a brake on the Moon's rotation, locking it into a synchronous spin rate – i.e. the Moon was brought to rotate on its axis in exactly the same time that it takes to circle the Earth, thus constantly turning the same side toward our planet. This 'synching up' of the Earth–Moon system occured early on in the history of the couple, at the time of the magma ocean when our hot satellite was undergoing crust/mantle differentiation. The Moon's denser layers might have bulged Earth-wise in the process, with the mantle rising closer to the surface under a stretched, thinner crust.

It is also possible that collision trajectories of asteroids or comets caused the late, large impacts to occur mostly on the near side of the Moon, both thinning and faulting the near side crust.

Regardless of the cause, if the dense mantle – source of the mare lavas – was closer to the surface on the near side than on the far side, this meant that lithospheric pressures on the upper mantle were slightly lower there, and the chances of melting increased for any given temperature. It also meant that once melting did start, the magma had less distance to travel to the surface, and a better chance to make it all the way up before cooling to a halt.

The second general trend of mare volcanism on the near side is its three-stage history: early, high-titanium, eastern quadrant basalts; global near side, low-titanium basalts; and late stage, western quadrant high-titanium basalts, with the activity level declining sharply past the 3 billion year mark.

According to recent Galileo data, mare outcrops embrace less of a chemical range on the limbs and the far side, perhaps because of mantle homogeneity there, or the influence of a thicker crust: in these regions, mare patches and cryptomaria are mostly low to medium-titanium basalts, with little regional variability.

A third, global characteristic of mare volcanism is its relatively limited life span, a direct consequence of the Moon's small size: whatever internal heat was accumulated through accretion and radioactive decay within the mantle was efficiently conducted to the surface and radiated to space, because of the Moon's advantageous surface-to-volume ratio (see Chapter 1).

Over the years, this efficient cooling lowered the Moon's upper mantle temperatures by several hundred degrees, and the frontier of partial melting migrated to ever-increasing depths. This level is now close to 1000 km down – a small, partially-molten asthenosphere, which embraces the last few hundred kilometers of mantle, down to the central core. It is not surprising, therefore, that any partial melt at such depths has virtually no chance of reaching the surface today, confined as it is by the thick overlying lithosphere. Fractures at such depths and pressures are self-annealing, and bar the passage of magma, although there is some evidence that volcanic gas can still seep to the surface when fractures come in to play, as we shall see in our last section.

The Moon does show a small amount of tectonic activity, as recorded by the Apollo seismometers: discrete moonquakes are focused at the base of the rigid lithosphere between 600 km and 1000 km depth; the total energy released is a mere

2×10^6 joules per year – one hundred billion times less than the seismic energy released annually by our own planet.

Significantly, these discrete moonquakes are bunched in pulses every two weeks, each time the Moon goes through the perihelia and aphelia points of its eccentric orbit around the Earth. In other words, moonquakes are due to tidal stresses exerted by the tug of our planet's gravity, rather than by any surviving, thermal mechanism at depth. From a magmatic point of view, the Moon is considered today to be clinically dead.

Lava flows on the Moon

When the Moon was volcanically active, lavas which erupted on the basin floors did so with great fluidity, and spread over wide areas before slowing to a halt. They owed their low viscosity to their high content of unobtrusive metal oxides (iron, magnesium, titanium), and correspondingly low silica content: less than 45% SiO_2 in average lunar basalts, and even down to 38% for Apollo 17's titanium-rich basalts.

By melting Apollo samples in the lab, petrologists were able to measure directly the viscosity of lunar basalts, and they obtained very low figures of 5 to 10 poises for a magma at $1200\,°C$. This is comparable to the low-viscosity komatiite lavas which were emplaced in the early history of the Earth. There is also some evidence that lunar lavas erupted at temperatures well above $1200\,°C$ – perhaps $1400\,°C$ to $1600\,°C$ – and their viscosity would have been lower yet as a result.

Another factor increased the flow range of lunar lavas: their high eruption rate. Large volumes of magma emprison their heat more efficiently than thinner sheets, and thus spread over larger distances before cooling. For all these reasons – high fluidity, temperature, and rate of eruption – lunar lavas flowed far and wide, on the slightest of slopes. As they circled the Moon, the Apollo astronauts were able to photograph extensive drape-like flows in the lunar maria, that showed up in subtle detail under the grazing sun – the most spectacular lying in Mare Imbrium. Shadows cast by flow fronts, combined with laser altimetry, indicated individual flows 40 to 50 m thick, with axial channels 1 to 2 km wide that were instrumental in funneling the hot lava.

How large were the mare eruptions, compared to volcanic events on Earth? We lack present day equivalents: even Iceland's Laki eruption of 1783 – the largest in recorded history – falls short of the estimated lunar flow rates. The Laki event took place along 30 km of fissure, and buried an area of 565 km² under 100 to 200 m of lava, in less than seven months. The eruptive rate was largest during the first two months, when it reached 0.1 km³ per day.

The Laki figures still fall short of the tremendous rates calculated for the lunar mare eruptions. A better equivalent on Earth is trap volcanism, where hot spot activity disgorges vast amounts of basalt. In the most recent trap event – which formed the Columbia River Plateau 15 million years ago – the main fissure was 130 km long, and fed the advancing flows at rates of 1 to 3 km³ a day, ten to thirty times the rate of the Laki eruption. This we can picture as a ribbon of hot lava all along the 130 km fissure, 10 m thick, flowing down the slopes at a clipping 500 to 1500 meters an hour – that is, covering 10 to 30 km in one day!

On the Moon, eruptive outputs were probably in the same range. In areal extent alone, most mare deposits are comparable to trap provinces on Earth (Mare Imbrium is $500\,000\,km^2$, the exposed area of the Deccan or Parana traps on Earth).

Individual mare flows are also as thick, and even thicker than those of terrestrial traps – a sign of high output, and a guarantee of long run-out distances. Crystal sizes in Apollo basalts indicate original flow thicknesses of around 10 m. At Hadley Rille, the Apollo 15 astronauts were able to photograph in cross-section a lava layer 15 to 20 m thick, jutting out of the rubble as a single flow. And at the high end of the range, shadow and altimetry measurements in Mare Imbrium place maximum flow thicknesses around 50 m.

These last figures are roughly double those on Earth: individual flows in the Columbia and Deccan traps average between 10 and 20 m, rarely 30 m. Why would lava flows be thicker on the Moon than on Earth? One reason could be wider eruptive vents, and larger discharge rates at the source. Alternatively, rocky levees on the margins of lava channels build up to greater heights on the Moon – because of its lesser gravity – and thus funnel thicker flows between their walls.

Be it as it may, thicker flows on the Moon partially explain the larger run-out distance: other conditions being equal, doubling the thickness of a lava flow amounts to multiplying its reach six-fold.

Whereas the longest lava flow on Earth (a trap unit in Australia) reaches 150 km, flows on the Moon are many times longer, as in the southwestern corner of Mare Imbrium, where three overlapping units are 1200 km, 600 km and 400 km long (the decreasing range pattern shows that the eruptive rate in Imbrium at the time was declining).

The flooding of the basins

Concerning global mare thickness, estimates can be drawn by using impact craters as markers: where young impacts blast through lava flows down to the underlying basement, the ejecta change color. The crater diameter at which this color change occurs indicates the thickness of the transpierced lava pile (from an empirical depth-to-diameter scaling law for craters). Alternatively, craters older than lava flows can also be used as markers: depending on how much their rims are surrounded, embayed, or breached by a lava pile, it is possible to guess the pile's thickness.

The total thickness of mare lava fill can also be estimated from the seismometry data collected by the Apollo science stations. An experiment in active seismometry was conducted during the Apollo 17 mission: astronauts Cernan and Schmitt planted eight explosive devices on the Taurus-Littrow valley floor, along with a network of geophones. After the astronauts left the Moon, the explosives were detonated by remote-control, and the echoes reverberated by the underlying strata were recorded by the geophones. From the data, it appeared that the total lava thickness reached 1200 m.

There is a last way to estimate the thickness of lava piles in the lunar basins. In 1968, the discovery was made that many circular basins on the Moon exerted a tug on orbiting spacecraft, as if dense rock layers below were spiking the local gravity field. The extra acceleration was slight indeed – satellites picked up a few millimeters per second of speed – but it was enough to signal significant mass concentrations beneath those basins, which were named *mascons*.

Geologists speculated at first that the extra mass represented leftover meteoritic material from the blasts that excavated the basins – iron 'shrapnel' buried at depth. But finer modeling of the anomalies soon identified their source to be the sheets of lava themselves, which filled the basins with denser rock. Gravity anomalies could then be used to estimate mare thicknesses: 3500 m of lava were calculated for the central areas of the Sea of Serenity; and 5000 to 6000 m for most of the Mare Imbrium, with central pockets up to 8000 m deep.

Other basins with few or no mascons were believed to have much thinner mare flows: the Sea of Tranquillity and the Ocean of Storms probably total a few hundred meters in global thickness, rather than the thousands of meters that characterize the pronounced mascon basins (another possibility is that older basins like Tranquillity are isostatically compensated, and no longer display marked gravity anomalies).

If we integrate these estimates over the entire surface of the lunar maria, we can derive an average figure for the total volume of lavas erupted in the course of mare volcanism: approximately six million cubic kilometers, which is comparable to the volume of the Deccan, Parana, and Siberian traps combined.

In search of volcanoes

Despite this remarkable output, there are no prominent lava shields on the Moon, and no towering stratocones built of layered flows of lava and ash. Only a few hundred small shields and domes, typically four to five kilometers in diameter and a few hundred meters in height, qualify as volcano candidates, and have yet to be visited.

There seem to be three major reasons for the absence of prominent volcanoes on the Moon.

Firstly, lunar volcanism is limited to basaltic chemistries. There is no significant andesite, trachyte, dacite, or rhyolite on the Moon, and therefore none of the viscous flows that stack up to build steep features, as is current on Earth. The fluidity of lunar lavas is extreme.

The second reason for the lack of relief is the discharge rate of lunar eruptions. On Earth, even fluid lavas can build up noticeable shields if they are extruded at modest eruption rates: small flows cool fast, and therefore congeal close to the vent, stacking up to form a relief – a shield, or even a steeper cone if the discharge is choked to a trickle. On the Moon, voluminous flows were the rule – because of pervasive upwelling – and spread over large areas and down the slightest of slopes, to create the subdued, lava plain topography we witness today. Whatever fissures were responsible for the eruptions were covered up by their own flows or neighboring ones, and literally vanished in the field.

There is a third reason for the lack of volcanic features on the Moon. On Earth, eruptions can build spatter cones, where clumps of magma tossed from the vent are strapped down by gravity and air resistance, and stack up as lava mounds. Finer ash also falls back in the immediate vicinity of eruptive vents, to build cones. But the much lower gravity on the Moon, and the lack of aerodynamic braking of the airless environment, let lunar ejecta fly far and wide: instead of agglutinating into steep features around the vents, lunar projectiles spread out into very thin disks.

Calculations show that on the Moon, only the largest clumps of lava will fall back around a vent, and lunar spatter cones will be limited to a few dozen meters in width. As for the finer droplets of magma – in the millimeter range – they will be propelled by the eruptive gases dozens of kilometers away from the vent: spread out so thinly, volcanic ejecta on the Moon are barred from building any significant relief.

Dark-colored disks, centered around small pits, have been described in many areas of the Moon – especially on the edge of the mare basins. Named *dark-halo craters*, some are believed to be pyroclastic vents, surrounded by their shallow, far-flung deposits – typically 2 to 10 km wide. One such crater was visited by the Apollo 17 astronauts in the Taurus-Littrow lava plains, but turned out to be an impact crater that churned up dark layers of underlying pyroclastics. However, other dark-halo craters might well be volcanic in origin – in particular those located along rilles and fissures.

Remote-sensing from orbit has shown these Dark Mantle Deposits (DMD for short) to include both reddish, iron-bearing glasses (as on the Aristarchus plateau), and high-titanium, blue deposits (as in the Taurus-Littrow valley). Data from the Galileo and Clementine probes indicates that these pyroclastic deposits are more widespread than was once thought, and include abundant patches in the arctic region of the Moon that has only recently come under scrutiny.

Shields and domes on the Moon

Other features qualify as volcanic candidates on the Moon. Heading the list of would-be volcanoes are the Marius Hills, a prime landing site for the last Apollo missions, before the program was curtailed. Grouped in the center of the Ocean of Storms, these twenty-odd shields and domes range 4 to 17 km in width, and rise 100 to 250 m above the mare plains. Some have summit depressions believed to be pit craters; and the narrow flows around their base are reminiscent of viscous andesite and trachyte flows on Earth.

For these reasons, the Marius Hills could be among the few examples of differentiated magmatism on the Moon, a differentiation that took place in waterless conditions – perhaps as the last dregs of a crystallizing magma reservoir erupted late in a mare sequence. Alternatively, or in addition, low eruptive rates might have limited the extent of the flows, building up the relief.

We get the same impression of differentiated volcanism at Rumker Hills, 1000 km to the north. This volcanic complex, on the western flank of the Aristarchus plateau, shows up as a broad swell, 140 km in width, built of overlapping flows, and capped by

FIG. 4.9. The Aristarchus plateau in the Ocean of Storms. Photographed by Al Worden (Apollo 15), the view looks south across the rugged hills: craters Aristarchus (left) and Herodotus (right) dominate the plateau, with Schröter's Valley (a graben occupied by a lava channel) emerging from under the rim of Herodotus and heading lower frame right. Herodotus is older than the mare flows and is flooded by lava, whereas Aristarchus impacted later and superimposed its ejecta on the Ocean of Storms. The mare fill is believed to be thin – a few hundred meters – as revealed by numerous ghost craters still protruding above the lavas (notice the ring above and to the left of Aristarchus). Credit: NASA.

four, gently-sloping domes. Several smaller, steeper domes surround the complex in the mare plains.

Halfway between Marius Hills and Rumker Hills, the Aristarchus plateau itself is perhaps the most spectacular volcanic province on the Moon (see fig. 4.9). A very large lava channel – known as Schröter's Valley – emerges from the flank of lava-flooded Herodotus crater, and snakes its way towards the plains. Because of the tadpole-shaped cleft at its source, the channel complex is familiarly known as the Cobra Head. The Aristarchus plateau also includes a number of other sinuous rilles, and a variety of pit crater chains, low shields, and lava flows.

Completing the volcanic line-up through the Ocean of Storms, Harbinger Mountains and the Gruithuisen Domes stretch out to the northeast of the Aristarchus plateau, into the fault-controlled cross-over between the Ocean of Storms and Mare Imbrium. The mare contact with the older highlands in this interesting area shows evidence of normal faulting, perhaps providing the setting for a pulse of differentiated volcanism: the Gruithuisen Domes are steep (1500 m in height for a 15 km base), and might be composed of aluminous, feldspar-rich magma.

All these regions are limited in area, but show an interesting trend in their distribution: Rumker, Aristarchus, and Marius Hills all line up roughly north–south in the Ocean of Storms, along with a subdued formation known as the Flamsteed Ring, and interpreted by some to be an igneous, arcuate protrusion beneath the mare lavas. The Gruithuisen and Harbinger Domes are close by, forming a subparallel

FIG. 4.10. Steep-sided mounds protruding from the mare flows on the near side of the Moon (photo by Ken Mattingly, Apollo 16). Although of unknown origin, these mounds could be extrusive domes of volcanic origin. Note the lava flows extending from the extremity of the mounds (lower left and upper right). Credit: NASA.

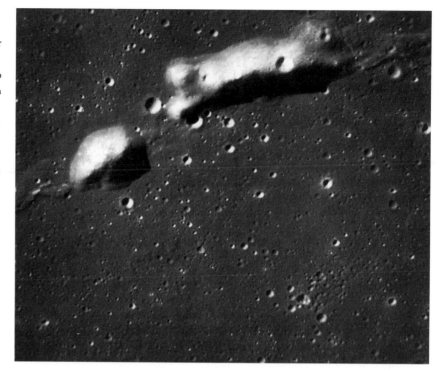

trend to the northeast. Some form of fault control must be responsible for the line-up, but it does take a leap of faith to attribute the sequence of volcanic centers to an interconnected string of upwellings in the lunar mantle, as some have suggested.

Besides these domes, which might point to independent pods of differentiated magma, a number of subdued shields are also found on the edge of the maria – several hundred in all, most visible on the orbital photographs of Apollo 16 and 17, taken under grazing sunlight conditions. Averaging 10 km in diameter and only a couple of hundred meters in height, these discrete shields are most likely the result of a drop in magma discharge rates in the final stages of mare flooding, the shortened flows building up around the vents. In essence, these features are the lunar equivalent of plains shields on Earth, as in Idaho's Snake River Plain (see Chapter 2).

In summary, the Moon totals over 400 domes, and at least as many small shields, located for the most part on the margins of mare-flooded basins, where fault control is prominent.

Lastly, it should be pointed out that a few large craters cannot be ruled out as possible volcanic calderas, rather than impact astroblemes. Herodotus crater on the Aristarchus plateau, towering over the Cobra Head channel, could be such a caldera. In Mare Orientale, on the lunar far side, a young crater 35 km in diameter has a smooth floor and lacks the blanket of ejecta characteristic of impact. The Sabine craters in the Sea of Tranquillity, and a number of other circular features round off the list of suspects.

Hints of recent volcanism on the Moon

To further boost the hopes of volcano fans, the possibility of some discrete activity continuing to this day cannot be completely ruled out. Over the aeons, the zone of partial melting has migrated to great depths inside the Moon – down to 1000 km – so that magma is no longer expected to reach the surface. But the release of built-up gas at depth is still a possibility, especially in view of the tidal stresses that the Moon experiences twice a month on its orbit around the Earth – at periapsis and apoapsis.

Indeed, if no on-going volcanic manifestations were ever spotted on space probe imagery, nor signaled by the orbiting Apollo astronauts, there have been some other hints of activity. Remote-sensing instrumentation aboard the Apollo orbiters did detect small quantities of radon-222 gas (a decay product of uranium) over the faulted edges of the maria, as well as over craters Aristarchus and Grimaldi.

These data bring some credibility to the observations of several astronomers who, over the years, reported visible manifestations of gas or ash venting on the Moon. These 'lunar transient phenomena' – or TLPs as they came to be known – include two well documented events: soviet astronomer Kozyrev obtained a photographic spectrum of a cloud of gas over crater Alphonsus on November 19, 1958; and several astronomers at the Flagstaff observatory observed glowing, red spots on two occasions in 1963, above the Aristarchus plateau. The reported lunar transient phenomena occur mostly around periapsis, as do the radon peaks and minor seismic activity recorded by the Apollo instruments, supporting the theory that they are pulses of gas, released during tidal relaxation of deep-seated faults.

In fact, there are several patches of bright surface material in the lunar *maria* that might well represent sulfurous coatings deposited from recent gas eruptions. One such example is a peculiar, D-shaped caldera, 3 km across, photographed by Apollo 15 astronaut Al Worden north of the Sea of Vapors. It contains a patchwork of irregular domes, intertwined with very bright markings. A larger, equally intriguing splash of bright material is found in the Ocean of Storms: the Rainer Gamma Formation, 150 km long and 30 km wide. Other albedo markings of suspected volcanic origin occur east of Mare Smythii.

So there is always hope for those who would like to witness a contemporary volcanic eruption on the Moon, or at least find the trace of very recent activity. But apart from gas exhalations, chances are very slim indeed: to all intents and purposes, the Moon volcanically shut down two billion years ago – a direct result, as we saw, of its small size – marking the 'half-time' of the Solar System.

Thus, the Moon presents us with an unspoiled volcanic record of the early days of planetary genesis, stratification, and cooling. In Chapter 10, we will attempt to put this short but fascinating life into perspective, comparing it with the volcanic records of the Earth and other terrestrial planets.

5 The discovery of Mars

The red planet

Mars is the fourth planet away from the Sun, after Mercury, Venus, and the Earth. It circles outside our planet on a highly elliptical orbit which it covers in 687 days, swinging inwards to a perihelion 206 million kilometers from the Sun at closest approach, and moving out to 249 million kilometers at aphelion.

On our inner, faster orbit, we overtake Mars every 26 months, and in the most favorable oppositions come within 55 million kilometers of our neighbor. Mars appears to us as a red star – brightest at opposition – a color which caused it to be named after the war gods of the classical era: Ares to the Greeks; Mars to the Romans.

With the advent of telescopes, Mars enjoyed an ever rising popularity: its

FIG. 5.1. Photomosaic of Mars, assembled from over one hundred Viking 1 images. This is the view an astronaut would have of Mars from an altitude of 2500 km over the equator at 80° W. The 3000 km-long tectonic rift of Valles Marineris parallels the equator, opening up to the east on the dark limb region of chasma terrain. The Viking 1 landing site is on the horizon at 2 o'clock (bright region of Xanthe-Chryse). On the opposite limb to the left, two circular spots mark the emplacement of the easternmost Tharsis volcanoes: Pavonis Mons (in line with Valles Marineris) and Ascraeus Mons (above). Credit: NASA/JPL, courtesy of Dr Philippe Masson, photothèque planétaire d'Orsay.

trembling disk projected mysterious markings that delighted astronomers, and led to imaginative speculations. The irregular, dark spots and blurry lineaments were soon made out to be 'oases' and 'canals', and together with the very real vision of ice caps at the poles, turned Mars into a fragile, doomed abode of life, with intelligent beings struggling for survival, and digging canals to irrigate their parched, equatorial deserts.

The Mariner's odyssey

The hopes of finding advanced forms of life on Mars, however, were soon dashed: Earth-based spectra of the martian atmosphere revealed icy temperatures and very low gas pressures – chiefly carbon dioxide. The first flybys of Mars by space probes in the middle and late 1960s revealed a desolate, Moon-like landscape, obliterated by impact craters. Shifting color markings across the red planet – once thought to be the mark of vegetation cycles – turned out to be dust streaks seasonally remobilized by the thin atmosphere's very high winds.

It was such a dust storm – of planetary proportions – that Mariner 9 encountered on November 14, 1971, when it fired its retrorocket to become the first space probe to achieve martian orbit. Mariner 9 beamed back early photographs that were totally featureless, so intense was the dust shroud that enveloped the planet. After the limited imagery of earlier flybys, the first global reconnaissance mission was off to a disappointing start. But Mariner 9 could wait – it was parked in orbit – and the wait soon grew into riveting suspense for planetary geologists.

As the winds abated and the dust began to settle, four intriguing dark spots emerged from the cloud tops, progressively coming into focus as large crater-like arenas, 40 to 80 km in diameter. Were these ordinary impact craters, like those already spotted by the previous probes? Mission specialists came up with an exciting alternative: since they were the first features to pierce through the settling dust, they had to crown topographic highs. And what were large circular craters possibly doing at the top of the highest mountains on Mars, if they were not...volcanic calderas?

Indeed they were. Over the following weeks, the dust veil dropped to the ground, and the four giant volcanoes appeared in all their glory. The group straddled the equator, between 100° W and 140° W. Three features lined up on a NE/SW axis: they were christened Arsia, Pavonis, and Ascraeus Montes, from north to south. The fourth feature, detached to the northwest, corresponded to a familiar bright spot known to Earth-based astronomers as Nix Olympica, 'the snows of Olympus'. Renamed Olympus Mons, the shield appeared even broader and taller than its three companions (see fig. 5.2).

Giant volcanoes

The 'snows of Olympus' was a fitting name: altimetry readings by the Mariner spacecraft indicated that the summits peaked 20 000 m above the surrounding plateau – 27 000 m above the reference level on Mars. Perhaps sometime in the past the giant volcanoes were covered with ice and snow – for there is evidence that Mars

FIG. 5.2. Olympus Mons on Mars is the tallest volcano in the Solar System: the giant shield is 500 km by 600 km and towers 27 000 m above the reference level. This oblique view was taken by the Viking 1 Orbiter, looking to the southwest. Clouds of ice crystals crown the upper slopes. The complex summit caldera is 80 km wide: its scalloped outline is due to multiple cycles of magma withdrawal and chamber collapse. The most recent pit intersects the main ring of the caldera to the right (northwestern pit). Outside the caldera, two impact craters are visible at 3 and 7 o'clock on the volcano's upper slopes, with radial lava flows girdling the slope-break beneath them. Credit: NASA/JPL.

had a denser, water-rich atmosphere in the past. But today, the red planet has only a very thin atmosphere of carbon dioxide, with only minor traces of water vapor: low pressures of 6 hectopascals at ground level – and less at higher elevations – are insufficient to foster precipitation. Only at dawn is there a form of tenuous precipitation, when tiny crystals of carbon dioxide are swept to the ground by settling dust, and coat the ground with an ephemeral film of carbonic frost.

Thin cirrus clouds also form in the lee of volcanoes, stretching westward from the four summits in the early morning hours (see fig. 5.2). These patches of airborne ice crystals make up the bright spots familiar to Earth-based astronomers.

The four giant volcanoes of Mars are as large as they are tall. Olympus Mons has a diameter at the base of 550 to 600 km (the shield is slightly oblong in the NW/SE direction). Close to 500 000 km^2 in area – if one includes the basal lava plains – the volcanic complex is the size of a country like Spain. By comparison, the largest volcano on Earth is Mauna Loa in the Hawaiian Islands: measured on the sea floor, the shield's diameter averages 150 km, yielding an area only one thirtieth that of Olympus Mons.

The three companions of Olympus – Arsia, Pavonis, and Ascraeus Montes – are nearly as large, and just as tall. All boast similar profiles, which are characteristic of shield volcanoes: their slopes average 5° or less, and are built of fluid lava flows – individual flow fronts and channels can be resolved on the Mariner imagery. Like many of their terrestrial cousins, the martian shields have down-dropped summit calderas – the 'craters' that first appeared through the settling dust storm.

Are any of these giant volcanoes still active today? And if they no longer are, how recent were their last eruptions?

The four Tharsis giants have crisp-looking features, although a number of small impact craters are peppered across the shields, attesting to a venerable age. From the number of craters counted on each shield, one can derive age estimates, but the numbers are highly speculative, dependent as they are on the frequency of impacts on Mars – a rate which is poorly constrained. According to the youngest proposed scale, the surface of Arsia Mons is 700 million years old; that of Pavonis Mons 300 my; and that of Ascraeus Mons 100 my old. As for Olympus Mons, the upper part of the shield's surface could be as young as 30 my!

If these 'optimistic' estimates are correct, it would appear that the last eruptions of Olympus Mons, and even Ascraeus Mons, are extremely recent by planetary standards. One hundred million years equate to only 2% of the life of a planet (4.5 by), and it is then reasonable to hope that some form of limited activity might still occur to this day: how ironic it would be, if the great martian volcanoes had erupted over 98% of the planet's history, only to shut down just before the coming of man!

A guided tour of planet Mars

There is much more to Mars than four giant shields. Mariner 9 and the later Viking probes collected a data base of over 58 000 images of the surface, with resolutions down to 10 m: the red planet turns out to be a geologist's paradise, with a rich history of impact cratering, volcanism, tectonics, and erosive landscaping.

To get an idea of the fascinating diversity of planet Mars, one ideally should take a window seat aboard an orbiting spacecraft. Let us then imagine what it would be like to spin around the red world in eighty minutes.

On an equatorial orbit, high enough to embrace a wide ground track, we would begin our journey at the 180° meridian, where our impression – if we look to the north and south – is that Mars exhibits a profound asymmetry between two distinct hemispheres. South of the equator are heavily cratered highlands, not unlike those of the Moon. But if we look to the north, the land drops to rolling plains, with far fewer impact craters, and displays instead a variety of rounded, knobby hills; and dust streaks etching the surface. The boundary between the two provinces is a scalloped cliff front, which shows every sign of being fault-controlled: apparently something happened to the martian crust on a global scale to focus activity in the northern province, while the southern plateau virtually ceased to evolve.

As our trajectory carries us eastward along the equator, we might notice discrete volcanic features in the plains, under the grazing illumination of the morning sun – slight undulations and arcuate scarps that resemble the flow fronts and wrinkle ridges of the lunar maria.

Not only is the relief subdued in the martian plains, but on a finer scale the ground itself appears to be exceptionally smooth in places: radar probing turns up such low reflectance values in the plains west of Tharsis that the area has been nicknamed the 'stealth' zone. The plains are so radar-absorbing that they must be covered by a blanket of clay-sized particles meters in thickness, semi-porous, with no protruding boulders or even pebbles. There is a distinct possibility that these are fields of volcanic ash, spread on the leeward side of the Tharsis volcanoes.

The plains progressively slope up, from zero elevation at the start of our round to over 2000 m as we cross the 150° meridian. Minutes later, if we take a new altimetry reading one thousand kilometers farther to the east, the plains reach 3000 m. The slope is gentle: one or two tenths of a degree, which is comparable to the continental slope of our ocean floors on Earth. On Mars, this gentle slope nevertheless points to a major geographical boundary: the edge of the Tharsis uplift.

Crossing the Tharsis uplift

A glance to the port side of the spacecraft helps us take our bearings: the broad shield of Olympus Mons looms over the horizon, looking very much like some grounded volcanic island, with a steep base, rounded flanks, and thin cirrus clouds stretching leeward from the summit.

Below our spacecraft, the topography keeps rising as we cross the broad regional uplift, now distinctly volcanic in nature. By the time ground elevation reaches 5000 m, we spot fault lines and flow fronts, and two prominent shields fill our port window, speeding by at close range: first Biblis Patera, crowned by a wide circular caldera; then Ulysses Patera also showing an impressive caldera, its rim breached by two large impacts. Biblis and Ulysses look like very old volcanoes: not only are they stamped by impacts, but their lower slopes are encroached by much younger-looking plains.

The Tharsis uplift levels off at 9000 m, and its broad plateau serves as a pedestal for the giant volcanoes that tower over the region. Our trajectory takes us straight above the central shield of Pavonis Mons.

The slopes of Pavonis rise at an angle of about 5°, comparable to the incline of basaltic shields on Earth like the Galapagos, Réunion, and Hawaiian volcanoes. Concentric fault rings girdle the structure, criss-crossed by radial lava flows. Unlike the two small shields to the west, the very low density of impact craters suggests, for Pavonis, a surface age of under 300 million years. As we fly over the top, volcanic flows and fissures stand out sharply around the summit caldera – an arena 40 km in diameter, bounded by steep cliffs that drop 4000 m to the central floor. The caldera floor is relatively smooth, surfaced by fresh-looking lava flows with undulating ridges, typical of cooling and compression of massive volumes of lava.

Leaving the impressive caldera behind, we cross over the eastern flank, shield sloping down to merge with the Tharsis plateau.

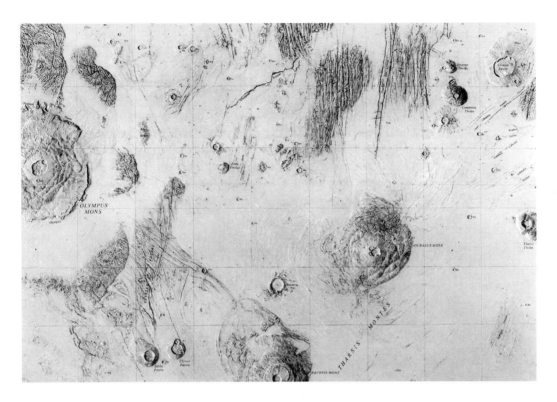

FIG. 5.3. Shaded relief map of the Tharsis quadrangle, showing a large section of the volcanic province. The map was prepared from Mariner 9 and Viking imagery. From west to east (photo references in parentheses): Olympus Mons (fig. 5.9): Biblis Patera and Ulysses Patera; Jovis Tholus; Pavonis Mons; Ascraeus Mons; Uranius Tholus; Ceraunius Tholus (fig. 6.5); Uranius Patera; and Tharsis Tholus (fig. 6.9). Credit: US Geological Survey.

Grand canyon on Mars

No sooner have we left the Tharsis volcanoes than a puzzling landscape captures our attention to the south. Puzzling is the word, for the plateau splits up into fault-bounded valleys and mesas, which give it the appearance of a giant mosaic, with its pieces pulled apart. Down in the graben, patches of mist hover at dawn in the shade of the steep cliffs: we have reached the head valleys of Noctis Labyrinthus – the labyrinth of night.

As we progress eastward, the labyrinth opens up into wider graben slicing open the Tharsis plateau. The landscape is out of proportion with anything we know on Earth: shadows on the floor of the main chasm indicate a 7000 m drop – three times the depth of the Grand Canyon. And where three parallel graben merge into one, the width of the giant rift exceeds 500 km. Baptised Valles Marineris, the structure runs 4000 km east–west, just south of the equator: transposed on Earth, the rift would stretch across the entire USA from coast to coast.

With its straight cliffs spreading at the base into triangular spurs, Valles Marineris is clearly tectonic in origin: extensional forces stretched the crust and down-dropped the central graben, a familiar process on Earth to which we owe the East African Rift; the Rhine graben in Europe; the Red Sea; and other graben of burgeoning oceans. Crustal stretching and thinning of this sort are usually accompanied by mantle uplift and volcanism, as exemplified by the shields of the East African Rift; or the Eifel volcanoes of the Rhine. Likewise, dark deposits on the floor of Valles Marineris, and distinct strata along the walls are interpreted to be volcanic flows, although volcanoes per se are lacking, and other processes – sedimentary in nature – could also account for the deposits.

Valles Marineris opens up to the east into a wide chaotic region that swings northward across the equator. We are only minutes away from the zero meridian – the halfway mark of our martian tour – and the late afternoon sun highlights the erosional features in the jumbled landscape that parades below us. Wide, sinuous channels groove the surface, like the streams of a large outwash basin: some channels even display braided courses, laminated floor deposits, sand bars, and tear-shaped islands, reminiscent of our flood plains on Earth. Although there are no intricate tributary systems characteristic of mature capture basins, these martian channels apparently were gouged by flowing water, during some catastrophic flooding event.

As the sun sets over the martian landscape, we see the channel scablands vanish to the north, replaced by the highlands spreading across the equator from the south to fill our windows. Back over dull, lunar-like terrain, perhaps we won't miss much excitement as we swing behind the dark side of the planet.

Our short orbital night might give us a few minutes to pause and ponder over martian mysteries: if the channels were indeed carved by running water, then there was a time on Mars when atmospheric pressures were higher and temperatures milder, so that water seeped up from underground reservoirs, and scoured the surface. Mars would then be the first planet we encounter that radically switched climate regimes in its history – from an early, warm and moist environment to a cold, dry, and virtually airless one. We can even speculate that the chemistry of life might have had an opportunity to blossom on the red planet, only to become frozen and irradiated out of existence when conditions deteriorated.

Volcanoes of Elysium

We are still spinning above cratered highlands as we emerge from darkness, but the end is in sight, plateau fronts receding south to make room for a new province of volcanic flows, encroaching from the north. On the horizon, we can make out the sloping bulge of another massive uplift, not quite as huge as Tharsis: the Elysium plateau. The volcanic rise is too far to the north for us to make out its features, but were we able to veer off course up to 30° N, we would fly over plains rising smoothly up to 5000 m, occasionally criss-crossed by arcuate graben and radial flow fronts, until we reached the top of the uplift, crowned by three volcanoes (see figs. 5.4 and 5.5).

The Elysium volcanoes are older than the Tharsis giants (they bear more impact craters), and their aspect points to a more varied range of volcanic activity.

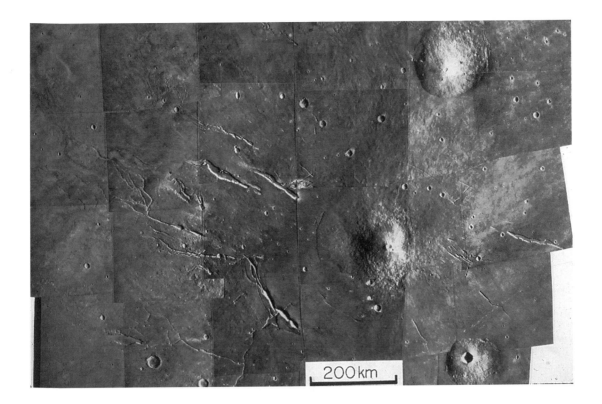

FIG. 5.4 The Elysium volcanic province with its three edifices: Hecates Tholus (upper right); Elysium Mons (right of center); and Albor Tholus (lower right, next to scale). Note the northwest-trending graben in the left half of the mosaic. Both Hecates Tholus and Elysium Mons have smooth upper flanks around their narrow summit craters, interpreted to be blankets of pyroclastic deposits. Compare lower half of mosaic with map (fig. 5.5). Credit: NASA photomosaic, courtesy of Brown University.

The southern-most feature, Albor Tholus, is a stubby shield with a wide caldera. Encroaching to the north, Elysium Mons is the largest of the trio, with a base diameter of 170 km, and a rounded summit towering 10 000 m above the plateau (15 000 m above the plains). Elysium has a summit caldera (10 km wide), and its slopes are relatively smooth, perhaps covered with ash.

Northern-most Hecates Tholus is about as large as Elysium Mons, but somewhat steeper, with finely fluted slopes, as if its surface was an eroded ash blanket rather than a hard lid of basalt. Could the narrow channels gouging the shield be ashflow corridors – common features on explosive volcanoes on Earth? Or else mudflow gullies from the catastrophic melt of water-saturated slopes, in warmer days of martian history? Chaotic deposits in the surrounding lowlands confirm this impression of mudflows, and add to the mystery of past climates on Mars.

So we leave the province of Elysium with an enlarged perspective, as we close our loop around the red planet. The volcanic plains of Amazonis – smooth and draped with dust streaks – fill our windows for the remainder of our flight, as we swing around to the zero meridian, and to our starting point.

New types of volcanoes

We should complete this survey of Mars by a look at what goes on at higher latitudes.

The poles of Mars are fascinating in their own right – immaculate ice caps

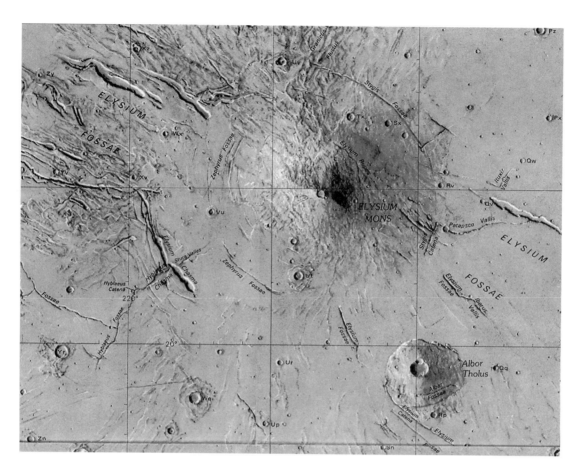

FIG. 5.5. Shaded relief map of the southern part of the Elysium volcanic province, prepared from Mariner 9 and Viking imagery. Compare with lower half of photomosaic (fig. 5.4). Note the circumferential graben around Elysium Mons, probably linked to the subsidence of the volcanic pile. Credit: US Geological Survey.

spreading over bright red terrain. The bulk of the ice sheets consists of frozen carbon dioxide – dry ice – which thaws away in the sunshine of early Spring, leaving behind inner, permanent cores of water ice. Around the ice sheets, benches of layered, wind-blown sediment – complete with dunes and terraces – record the seasonal cycles of sublimation and deposition of carbon dioxide, mixed with airborne dust.

In these high latitudes, the martian ground appears to behave plastically, in contrast to the more rigid-looking landscapes at the equator: impact craters have subdued rims, and finer details are smoothed out, as if erased by the slow creep of ice-saturated ground. The northern plains might well be the hiding place of the water that carved the channels and the flood plains in the early days of the planet's history. Conceivably, thick layers of permafrost could be trapped underground over hundreds of meters, from the poles down to the mid-latitudes, where the ground resumes a more rigid behavior.

Volcanism is also prevalent in the high northern latitudes, mostly in the form of subdued lava plains, peppered with small shields at the limit of our photographic resolution – lava and cinder cones punched with pit craters reminiscent of Icelandic volcanoes (see fig. 6.3). There is even one fluted feature the size of Mount Etna, the first stratovolcano-looking feature identified so far on a planet other than Earth. But

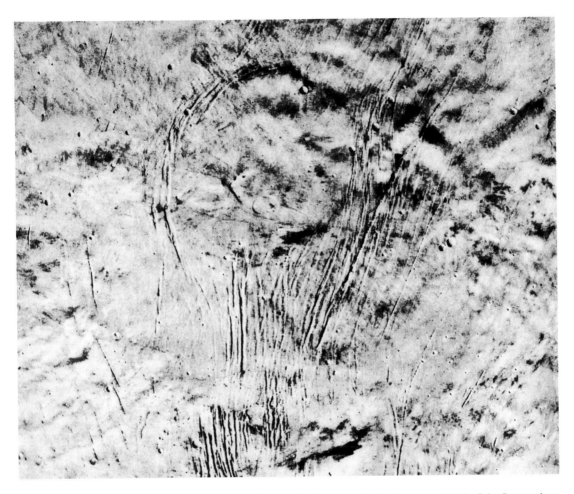

FIG. 5.6. With the planform of a giant light bulb, Alba Patera is the largest single volcanic construct in the Solar System. A set of peripheral faults girdles a central arena 700 km × 500 km, with a relief of only 2000 m. The boomerang-shaped dark 'filament' of the light bulb, left of center, is the western scarp of Alba's subdued main caldera. An inner, more rounded caldera is nested southeast of it. Lava flows cover the arena, and extend far and wide outside the graben, giving the volcanic pile an overall diameter of 1500 km. Most of the volcanic activity that built Alba Patera took place between 3 and 2 billion years ago. Credit: Mariner 9 photomosaic, from T. Mutch *et al.*, 1976, *The Geology of Mars*, Princeton University Press.

our biggest surprise comes from giant features, the likes of which we have never seen before, as we survey the temperate zones of the mid-latitudes.

Due north of the Tharsis plateau, an arena of lavas 600 km wide spreads across the plains, bounded by prominent, arcuate faults, and slightly convex in shape like a huge lens (see fig. 5.6). The lava disk rises very progressively from zero level on the periphery, to 3000 m at the rounded apex of the bulge. In the center of this broad shield, a subdued pair of interlaced calderas stare out like the pupil of some gigantic, cyclopean eye. On the outside, past the ring faults, sloping lava plains extend the overall dimensions of the volcanic construct to a record 1500 km.

Because of its unique profile, the volcano received an original type name: it was baptised a *patera* – the Greek name for saucer, so much the flattened shield resembles a shallow, overturned saucer.

Alba Patera, as it came to be known, is so huge that a traverse on foot would take an astronaut close to three months, at an average pace of 20 km a day. Such a trek would probably take much longer, however, for lava ridges and flow fronts would considerably hamper a hiker's progress, not to mention the huge, bounding graben, which slice through the ground like oversized highways, hundreds of kilometers long. One would have to cross dozens of these collapsed corridors before setting foot on Alba's central platform, and hike another 250 km to reach the central calderas – their broad floors peppered with small shields and domes, the size of cinder cones on Earth.

Alba Patera is a unique phenomenon, truly the only one of its kind in the entire Solar System. Other flattened volcanic constructs have been located on Mars, although much smaller in size: they have also been logged as paterae, on account of their unusual shape.

One subdued caldera with an apron of lava sits not far below the equator, south of Elysium: Apollinaris Patera. But the main cluster of paterae occurs farther south, around the Hellas impact basin, where three old and eroded features have been identified: Tyrrhena, Hadriaca, and Amphitrites Paterae. Each one of these volcanic aprons stamps the cratered crust with a wide, lava-coated arena, surrounded by large channels radiating in all directions like the spokes of giant bicycle wheels.

Tyrrhena Patera (see fig. 5.7) has been nicknamed 'Dandelion' on account of its petal-like pattern. Hadriaca Patera, immediately to the southwest (see fig. 6.7) is even more reminiscent of a flower, with its dense array of wavy channels. These highland patera attest to a very specific type of volcanism early in martian history, the density of impact craters pointing to ages in excess of 3 billion years, and probably closer to 4 billion years – a period known as the Noachian age, when 'wet' atmospheric conditions formed the presumed flood features that we noticed during our orbital tour. The prominent channels gouging the paterae might then be water-related, carved by the underground sapping of soft volcanic ash.

A field trip to Olympus Mons

If we summarize the Mariner findings, Mars boasts some twenty large volcanoes, and many smaller cones and shields, at the limit of our photographic resolution – nested in and around the major features, or sprinkled throughout the northern plains.

Of the twenty large volcanoes, half a dozen are the saucer-shaped paterae: four ancient features in the south, and two giant ones in the north – Alba Patera and a subdued depression known as Tempe Patera. The other dozen large features fall in the familiar category of lava shields, with the giant Olympus Mons heading the list.

Some of the shields are visible in their totality, like Arsia, Pavonis, and Ascraeus Mons, while others are embayed by younger volcanic flows burying their base, with only their upper, steeper flanks emerging. The shorter, stubbier appearance of the

FIG. 5.7. Tyrrhena Patera is one of the few volcanoes in Mars' southern hemisphere. It is also one of the oldest: depending on the impact crater scale used, the low-relief edifice is 2.3 to 3.7 billion years old. The feature has been nicknamed 'Dandelion' on account of its radial, petal-like pattern of erosive channels. The channel network is attributed to the release of volatiles (probably water) from porous ash layers, leading to the sapping of the surface. Credit: NASA, courtesy of the National Space Science Data Center, through the World Data Center-A for Rockets and Satellites (Viking Experiment Team Leader: Dr Michael H. Carr).

emerging summit has often led to a different type name – *tholus*, or 'hill' in Latin – but these features can best be described as embayed shield summits.

Apart from lava plains, shield building thus appears to be the dominant form of volcanism on Mars. From examples on Earth, we know that shields grow from the piling up of fluid lavas radially around a central magma chimney or rift zone – with lava emplacement occurring mostly through underground tubes or tunnels. Martian features, however, stand out on account of their sheer size. Olympus Mons, the giant, best illustrates the similarities and differences with terrestrial shields (see fig. 5.8): perhaps it will receive one day the visit of remote-controlled rovers, or a manned expedition to probe its rich volcanic history.

Approaching Olympus Mons from the air would be a visual enchantment to future astronauts. Before reaching the broad shield perched on the sloping horizon of Tharsis, a spacecraft coming in from the northwest would take its crew over a unique, jumbled terrain, the ground divided into a 'griddle-cake' pattern by criss-crossing ridges and grooves.

Nicknamed the aureole, the grooved terrain has been colorfully compared to the parched, dusty skin of an elephant, and its origin remains somewhat of a mystery. The formation is not only widespread in area – extending hundreds of thousands of square kilometers – but it is also substantially thick, perhaps up to a thousand meters in places, with arcuate lobe fronts and folds attesting to some form of dynamic,

FIG. 5.8. Olympus Mons, the giant volcano of Tharsis (north is upper right). The shield is 400 by 500 km wide and peaks at an altitude of 27 000 m. The lower part of the shield is truncated by a cliff – sheer in places (with a 6000 m drop), overrun in others by cascading lava flows and landslides (especially to the southwest and northeast). The upper part of the volcano is made up of a series of slope breaks and terraces, leading to a complex summit caldera 70 km in diameter. Credit: NASA/JPL, courtesy of Brown University.

flow-like emplacement. The most accepted theory is that the grooved apron is a series of huge landslide deposits, shed off the slopes of Olympus Mons onto the basal plains.

As our crew flies in closer to the shield, its attention would switch from the plains below to the broad silhouette now filling all of the flight deck windows, and displaying a characteristic 'bumpy' profile: the volcano rises as a stepwise succession of sloping ramps alternating with gentler terraces, giving it the undulating aspect of a Viking war shield. Even more startling is the jagged scarp line rimming the base of the mountain, a slope break overrun by cascading lava flows in places, and in others dropping nearly vertically – as on the northwestern flank, where our spacecraft would close in on a towering cliff face over 5000 m in height!

This would be a formidable obstacle for any expedition approaching Olympus Mons from the ground, albeit a breathtaking sight: along the north, northwest, and southeast sections, the cliffs of Olympus are three times the height of the Grand Canyon walls.

What process acted to remove such a thick section of lava and ash from the periphery of the shield, leaving a clear-cut scarp, with no sign of debris in its immediate vicinity? The vast aureole in the distal plains comes back to mind: perhaps the immense mass of Olympus Mons exerted such pressure on its lower flanks that an outside 'skirt' broke off along ring faults, and flowed out as a towering base surge, before settling in the plains as a thick blanket of debris.

The road to the top

If our astronauts manage to land their spacecraft at the top of the cliff, and disembark to continue on foot, their first impression would be that of a rolling plain, with few clues as to the uphill direction: the shield's lower slope starts off at a mere two to three degrees, so slight an incline that local flow fronts and gullies would mask the overall trend. As for the distant summit, it would lie out of sight, masked by the curvature of the horizon.

Taking its bearings from the Sun and stars, our crew would commence its 200 km trek to the summit caldera, an expedition that would take well over two weeks, perhaps even closer to a month if ridges and collapsed lava channels clog the way to the top. A far cry from the day or so that it takes to climb a terrestrial shield!

On Olympus Mons, the landscape would look unbearably sterile to those accustomed to the sugar cane fields and citrus groves of terrestrial volcanoes: not a shrub, not even a blade of grass on Mars to break the monotony of a mineral world frozen in time. The only motion might be the sifting of red dust between the jumbled rocks, funneled downslope by the wind.

The winds at the surface of Olympus Mons and other giant volcanoes are driven by the steep temperature and pressure gradients between the top and base, and tend to blow radially down or upslope. Temperature contrasts are stretched on Mars, where the atmosphere is too thin to convey heat efficiently and smooth out major differences: temperatures plummet to $-150\,°C$ at night, and jump to $0\,°C$ in the equatorial plains at noon – with the overprint of altitude gradients, which drive fast winds over the major reliefs.

From a geological perspective, the surface of Olympus Mons would not look too unfamiliar, except for the unusual vastness of its features. The rolling crests and troughs of lava flows running downslope fan out in subtle waves across the shield – arches 3 to 4 km wide, and 30 to 40 m in amplitude. Individual flows would be difficult to identify from the ground, although one might occasionally spot rocky lips on either side of a central groove – the leveed channel that streamed the molten lava downslope. On Olympus Mons, such axial channels are typically hundreds of meters wide: they once carried fluid magma over dozens of kilometers – either subaerially under Mars' open skies, or in tube fashion, under roofs of crusted-over basalt. Perhaps roof fragments still arch over channels in places, remnants of what were once long, twisting tunnels, before impacts and landslides shattered them to pieces.

Landslides are commonplace on Olympus Mons, as we can guess from the shield's marked slope breaks – reflecting constant readjustments in the balance of its massive flanks. Our exploring team would notice the first major change in slope after climbing the shield for a week or so, the incline suddenly doubling: from four or five degrees, the slope would jump to ten degrees over several kilometers, before returning to a more comfortable angle – a nearly horizontal terrace. Twenty to thirty kilometers farther, the slope again would steepen to lead up to the next terrace, giving Olympus Mons its characteristic, step-like profile.

We are used to spectacular slope breaks on terrestrial shields, namely on

Hawaiian volcanoes, but on closer examination the structures on Mars look different. At Mauna Loa and Kilauea, the shields are ringed by steep scarps named *pali*: these girdling faults accommodate failure of the mountain's massive flanks, which slip downslope like loose skirts along circumferential scarp planes.

On Olympus Mons however, the slope breaks are not as steep: they have a more rounded profile than the Hawaiian pali, pointing to a different tectonic regime. The present consensus is that the slope failures are due to downward shoves from the shield's overlying mass, rather than extensional tugs from the downslope part of the shield. Instead of normal fault slippage – the pali regime – the ground failure at Olympus Mons takes the form of thrust faulting from above, leading to the pile-up of failed layers as shallow-angle terraces.

A trek up the shield of Olympus might help resolve the issue, or then again it might not. But explorers would certainly welcome the last slope break with relief, the steeper final section leading to the top of the volcano: because summit eruptions are limited in volume, they tend to build stubbier flows and cap the mountain top with the steepest incline. Explorers, if they were lucky, might find a few open vents at the source of the terminal flows, half clogged with lava rubble and pockets of red sand. Patches of colorful sublimates might even coat the vents, streams of fine gas jetting forcefully into the near vacuum of the martian sky. Wishful thinking, perhaps. But the future might tell.

On the rim of the caldera

Trekking across the summit of Olympus Mons, it would take a day or two to reach the giant, central caldera. Deep fissures would increase in frequency before the ground suddenly dropped two thousand meters down to the floor of a giant amphitheater – a sheer cliff curving left and right, with no sign of the opposite rim on the horizon, so huge is the volcanic cauldron.

The caldera of Olympus Mons, 80 km in diameter, is a complex collapse structure – a frequent occurrence on volcanic shields, when the central magma column retreats downward, and withdraws its support from the summit. The resulting caldera is often resurfaced by lavas, when the magma column rises back to the top, and invests the peripheral faults to flood the crater floor. The Olympus caldera bears the mark of such resurfacing, with a smooth coating of lava, buckled up in places by wrinkle ridges. We can imagine a time – perhaps not too distant – when the giant caldera harbored a lava lake, glowing in the martian night.

Lava lakes are common occurrences in terrestrial calderas, although they are fairly small, in keeping with the craters themselves. Kilauea's lava lake in Hawaii played a major role in the pioneering days of volcanology, offering a permanent, natural laboratory for the study of magmas and eruptive processes. Magma filled the caldera pit for a century (1823 to 1924), before draining back down the volcano's plumbing system. Other famous, intermittent lava lakes include those at Nyiragongo, Nyamlagira, and Erta Ale in Africa; Piton de la Fournaise on the island of Réunion; and Mount Erebus in Antarctica.

FIG. 5.9. Topographic profile across Olympus Mons caldera. Insert shows cross-section represented (AA' line across the main caldera, looking to the northeast). Pits are numbered from youngest to oldest, 1 to 6. In the topographic profile, note extensional faults (graben) on the periphery of the caldera, and compressional ridges toward the center. Credit: Courtesy of Dr Thomas Watters, from 'Distribution in the Floor of the Olympus Mons Caldera', T. R. Watters and D. J. Chadwick, *LPSC* XXI, 1990.

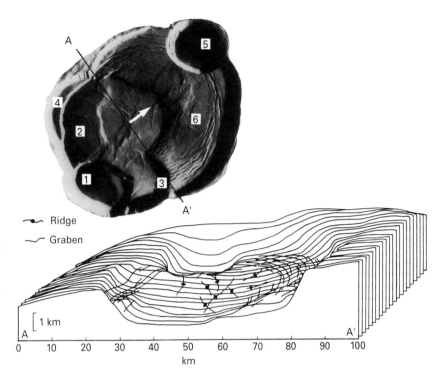

The Olympus Mons caldera is huge in proportion, and one is bewildered at the thought of its entire arena filling with lava, draining, and collapsing all in one unit, over hundreds of square kilometers. More realistically, activity at the caldera affected different areas at different times: intersecting floor levels attest to six separate stages of collapse, as magma pockets shifted below the summit. The caldera apparently started off as one wide arena, before activity migrated to its northeastern corner, where a deeper pit harbored the activity. The magma chimney then shifted to the western side, where successive cycles of flooding and down-faulting scalloped the rim into sweeping cliffs.

Standing on the northwestern rim, explorers would embrace a peripheral crescent of lava plains at the base of the cliffs, racing to the horizon and to the next drop, down to the floor of a younger, inner caldera (pit no. 2 in fig. 5.9). The youngest pit of all (no. 1) is farther off to the south.

The down-dropping of entire floor sections, as magma columns drain and their roofs collapse, tells us something about the shape and size of these shifting pockets of magma under the volcano's summit – known as magma chambers. Indeed, when a chamber fills with magma and expands, it stretches the surface of the shield. During an eruption, when a chamber funnels out the magma through fissures and vents, the volcano deflates, and its flanks relax.

This volcanic 'breathing' is measured on Earth through the use of tiltmeters, which record slope changes at checkpoints around the volcano – a great way to predict and monitor eruptions. Around Kilauea, slope changes as small as a few microradians are measured – less than one thousandth of a degree – which corresponds to a bulging or deflating of a few tens of centimeters over an area the size of a major city. Slim as they are, these ground swells, combined with the fine seismic tomography now available, make it possible to sketch 3-D images of underground magma chambers.

At Olympus Mons, activity might well have ceased, and tiltmeters would remain silent, but there are other ways of locating magma chambers. One is to look at the pattern of faults at the surface, and model the position and shape of the stress-creating chamber at depth. In the case of Olympus Mons, the bottom of the caldera shows a pattern of extensional faults at the periphery, giving way to compressional ridges 10 to 15 km into the arena, some 25 km from the center (see fig. 5.9). This can be modeled as representing the stress field caused by pressure change in a lens-shaped chamber, located 10 to 15 km below the summit.

The Olympus magma chamber would then be about level with the middle part of the shield. This happens to be the level where the prominent slope breaks and terraces occur. Therefore, it may well have been the cyclic expansion of the magma chamber which brought about bulging in this section of the shield, triggered slippage, and compressed the layers downslope.

The location of the stress-causing magma chamber, 10 to 15 km below the summit, would also have caused the lower third of the shield to undergo extensional stress, and this is indeed what we see in the form of the Olympus cliff – the steep scarp hinting that the lower third of the volcano wasted away in response to the stress field.

Thus, there is good agreement between the magma chamber model and the position and nature of tectonic features on Olympus Mons.

Shield building on Mars

Second in size to Olympus Mons, Arsia Mons is 420 km in diameter, and rises 19 km above the Tharsis plateau, 27 km above the plains. Arsia distinguishes itself by a very large caldera, some 120 km in diameter – twice the area of the Olympus caldera. It is a simple, circular feature, down-dropped between concentric fault rings. Outside the caldera, circular graben are also found farther downslope, girdling the shield: they are the source of many a lava flow. Near the summit, two radial rift zones breach the concentric graben, in the northeast and southwest 'corners' – a popular tectonic direction on the Tharsis plateau (on a wider scale, Arsia, Pavonis, and Ascraeus Mons are themselves aligned on this axis).

Individual lava flows are visible at all elevations on Arsia Mons, coating relatively steep slopes near the top (over 3°), and shallow inclines (less than a degree) on the lower flanks. Differences in slope translate into differences in flow behavior. On the steeper upper flanks, channels are narrow – 300 m wide at best – and the fast-running flows are encased by rocky levees. Farther down shield, where the slopes are gentler, lava flows tend to spread out in thinner sheets, 4 to 5 km in span. And on the lower flanks, where the shield grades into the plateau, flow fronts fan out to widths of 40 km and more.

This difference in morphology is due primarily to slope, but also to discharge rate at the vent – two factors that are tightly interconnected. We know from volcanoes on Earth that short summit flows pile up and steepen the slope because eruptions there are short-lived, and limited in volume. Greater discharge rates which occur farther downslope – usually level with the magma chambers – lead to the wide, more voluminous flows, which run out over longer distances because they remain hot and fluid for longer periods of time. One can directly relate discharge rate to run-out distances, a relationship that is relatively independent of other flow parameters.

Using these equations for the Arsia flow, 300 km-long features indicate that the eruption rates were somewhere between 100 000 and 1 million cubic meters per second, comparable to the discharge rate of the Amazon River.

High rates of the kind are not unheard of on Earth, but they are rare and usually short-lived. The one historic example of comparable magnitude is the start-up discharge rate of the 1957 Hekla eruption in Iceland: 100 000 m^3/s during the first day, after which the flow rate dropped by over an order of magnitude for the remainder of the eruption. Comparable eruption rates, but this time on the order of weeks, are postulated for the great flows of the Columbia River Plateau and other traps. Eruptions on martian shields probably lasted weeks at a time as well. Layer after layer, the Tharsis shields took millions of years to reach their present size, alternating high discharge rates with long periods of rest.

Volcanic marathons

We come to the apparent paradox that the volcanoes of Tharsis are both very young and very old. They are very young on the outside, where the most recent eruptions have resurfaced the slopes, especially in areas close to the summit. And they are very old because the sheer volume of the lava pile represents hundreds of millions of years of intermittent activity – in some cases perhaps as much as 2 to 3 billion years (Olympus Mons, Alba Patera).

Such longevity and volume are exceptional. On Earth, our largest shields grow to completion on timescales of a few million years at best, before going extinct and falling prey to erosion. Mount Etna reached its present size in 400 000 years; and the great Mauna Loa/Kilauea shield in Hawaii built up over a million years, but it is nearing completion and will soon shut down.

What is the secret of martian longevity, that has led to the growth of such monstrous volcanoes? The answer has to do with convection regimes within the planet, and the fixed nature of the lithosphere. On Earth the convecting mantle breaks up the thin lithosphere above it, and jostles plates at the surface, that move relative to the deep mantle beneath. Volcanoes which grow on these plates are therefore severed from their magma sources quite rapidly, as the plates carry them away from the hot spots. Terrestrial volcanoes drift away from their mantle 'roots' at rates of several centimeters per year – several meters per century. After a million years or so of this regime, the offset is such between a volcano and its magma source that the volcano is disconnected and shuts down, while another feature builds up more directly over the hot spot. This 'smearing' of hot spot activity by moving plates on Earth is described in Chapter 2.

On Mars, it is the fixity of the lithosphere that contributes to the gigantic dimensions of volcanoes. A monster like Olympus Mons fixes in one place 2 to 3 billion years of hot spot activity, with the older lavas buried at the bottom of the shield (a great place to study them would be the basal scarp cutting through the layers), and the younger lavas emplaced on top, with the youngest near the summit caldera. This is where we should look for the freshest samples to date, in order to find out how recent the last eruptions on Olympus really were. Unfortunately, the upper slopes of Olympus are not a good target for landing missions: plains and basins are preferred, because their greater atmospheric pressure assists the aero-braking of landing spacecraft.

Landing on Mars is indeed our next step, if we want to collect ground truth to confirm and enrich our orbital surveys. Such was the goal of the spectacular Viking mission of 1976, that landed two probes in the middle of the martian lava fields – one of the most successful missions of planetary exploration.

6 Martian volcanism in space and time

Landfall on Mars

On July 29, 1976, the automatic landing craft Viking 1 disengaged from its orbital bus and fired its retrorockets to attempt the first successful landing on the red planet. Prior to module separation, the Viking bus had spent five weeks in martian orbit, surveying potential landing sites with its telephoto camera. The youngest volcanic regions, despite their great scientific appeal, had been discarded on the grounds of altitude or surface roughness, and the choice fell on the outwash basin of Chryse Planitia – ancient lava flows believed to have been eroded and resurfaced by water outbreaks in the early 'Noachian' days of martian history.

While the orbital bus stayed parked in orbit to relay its radio signals and continue mapping the red planet, the Viking 1 lander plummeted through the tenuous, upper layers of the atmosphere at 4 km/s, shielded from the heat of entry by a wok-shaped protective lid. The braking of the atmosphere slowed the craft down to 250 m/s by the time it crossed the 6000 m elevation mark, at which time the mission sequencer deployed a 15 m parachute, ejected the braking shield, and deployed three landing feet. One thousand meters from the ground, Viking discarded its chute and fired three sets of retrorockets, throttled by radar. Further decelerated by the impulse, the probe made contact with the ground at a gentle settling speed of 2 m/s.

With no time to waste, Viking 1 immediately powered up its two cameras, and scanned the first close-up picture of the surface of Mars, followed five minutes later by a wide-angle panorama of the Chryse landing site. The remarkably sharp black and white images unravelled a volcanic plain rolling out to the horizon, peppered with lava rocks and boulders, and 'sand' pockets grading to dunes several meters high in places – a landscape reminiscent of eroded lava flows of the American Southwest (see fig. 6.1).

Color images were also secured with a set of filters: as expected, the ground was orange-tinted, arguably because of the oxidation of iron-bearing minerals in the soil and rocks – suggestive of mafic lavas. Photographed in close-up, many martian rocks appeared riddled with small pits, resembling the vacuoles blown by escaping gases in the lava. Rocks at the Chryse landing site spanned all sizes, from small clods of soil to boulders the size of cars – including one prominent boulder 8 m from the lander, measuring 3 m $\times$ 1 m (extreme left, in fig. 6.2). Most rocks, however, averaged 10 to 50 cm in length.

Soil analyses

The Viking lander was fitted with a telescopic, robotic arm designed to test the mechanical properties of the ground, push around rocks, and collect soil samples for delivery to a set of onboard experiments. Samples collected by the robotic arm were

FIG. 6.1. The plains of Chryse (22° N, 48° W) photographed by Viking 1 on August 3, 1976. The view looks east against the morning sun and covers 100° from left to right (northeast to southeast). The plains appear to be lava fields, reworked by erosion (the large boulder at left, eight meters from the spacecraft, is three meters wide). The orientation of dunes indicates the dominant wind to blow from upper left to lower right of frame. Credit: NASA/JPL.

FIG. 6.2. Viking 2 landed in the plains of Utopia (44° N, 226° W), north of the Elysium volcanic province. The uniform size and distribution of boulders is typical of lava flow erosion, the landform is reminiscent of volcanic deserts on Earth. Most lava rocks are vesicular. The horizon is slightly slanted (one of the probe's footpads rests on a rock). Past the footpad at right notice the metallic instrument cover lying in the patch of light-colored sand. Credit: NASA/JPL.

dropped into a funnel-shaped opening in the roof of the spacecraft, and distributed to five test cells, each measuring six cubic centimeters – the size of cigarette lighters.

Four of the cells were designed to test martian soil for signs of biological activity. Much to the experimenters' chagrin, all test runs came up negative: some surprising chemical activity did take place, but apparently resulted from inorganic superoxides.

So great was the effort to detect life that geologists were only awarded one experiment cell aboard the Vikings – and a last minute concession at that. The device was a pocket-sized X-ray spectrometer, designed to analyze martian soil by irradiation: under the impulse of X-ray bombardment, the soil radiates energy at wavelengths characteristic of constituent atoms – a phenomenon known as X-ray fluorescence.

Rocks on the site were of course too large to be funneled down to the test cells; only clods of soil were accepted by the 12 mm sieve, which protected the delivery system. Hence, it was the composition of martian soil – not whole rock – that was measured for the first time on July 27, 1976.

Silica ranked first among the measured oxides (44%), and iron oxides came in second (17.5%), far ahead of aluminum (7%), magnesium (6%), and calcium (6%) oxides. Sulfur oxide also peaked at 7%, a much higher figure than in terrestrial rocks and soil. Minor elements on the Viking site also included a pinch of titanium oxide (0.5%), and chlorine (0.4 to 0.8%). The greatest surprise, apart from the abundant sulfur, came from the very low potassium content, well below the sensitivity threshold of the analyzer – scoring under 0.1%.

The abundance of iron oxide, and relatively low proportion of silica would tend to place martian soil in the broad category of basalts and associated mafic rocks, although the complete chemical profile (across all measured oxides) does not fit any specific rock type known on Earth. This was to be expected, first because Mars is a different planet (atomic proportions differ slightly from planet to planet), and also because the Viking samples are clods of soil – i.e. a mixture of mineral grains eroded from a variety of rocks, rather than representing a single rock type.

Additional test runs confirmed the early results, both on the Viking 1 Chryse site, and in the plains of Utopia where the twin spacecraft Viking 2 landed in early August. On both sites – more than 4000 km apart – X-ray fluorescence measured identical compositions, confirming that the reddish dust was an averaged-out mixture of eroded minerals, homogenized and distributed by the wind on a planetary scale.

Despite not being primary rock signatures, soil compositions tell us something about the nature of martian lavas. Firstly, the high ratios of iron and magnesium point to high-temperature, mafic magmas, which imply great fluidity. This is in keeping with the shallow slopes and long run-out distances of lava flows on the red planet.

Secondly, the low potassium content of the soil (below the threshold of detectability) indicates that martian magmas have not undergone substantial differentiation. Indeed, concentrations of alkali elements like potassium usually reflect the amount of 'distillation' that a magma undergoes through successive cycles

FIG. 6.3. Plains volcanism is believed to account for the vast majority of the lava flows extruded on Mars. This view of the Cydonia region (38° N, 12° W) is on the edge of the dichotomy that separates the volcanic northern provinces from the southern Arabia highlands. In the upper right corner, note the bright cones with elongate pit craters, several kilometers wide at the base. The large cone to the left is built on an array of fractures. The lower half of the image is peppered by a dense population of small bright cones, a few hundred meters in width, often crowned with pit craters. Credit: NASA, courtesy of the National Space Science Data Center, through the World Data Center-A for Rockets and Satellites (Viking Experiment Team Leader: Dr Michael H. Carr).

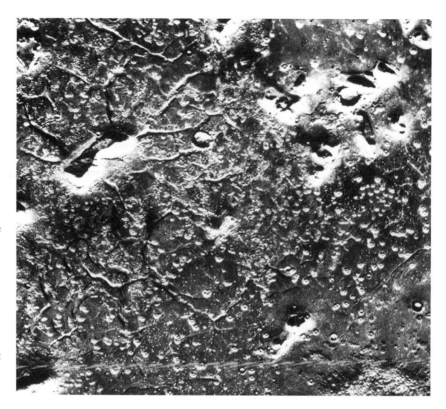

of congealing and remelting. Alkali enrichment also occurs when rising magma traverses and integrates potassium-rich crustal rocks – a process known as contamination. The near absence of potassium in martian soil establishes that neither process was dominant, and that magmas on the red planet were mantle melts, erupted directly to the surface with little chemical evolution along the way.

Plains volcanism on Mars

Chemical analyses are consistent with the appearance and extent of volcanic plains on Mars: lava plains cover close to 60% of the planet's surface – up to 80% in the northern hemisphere – and are thought to represent voluminous outflows of primitive, fluid magma, erupted at large discharge rates. Volcanic plains coat the vast uplifts of Tharsis and Elysium; fill giant basins like Chryse and Amazonis; and stretch over the vast northern plains of Vastitas Borealis. Lava plains are less widespread in the southern hemisphere, although ridged plains are noticeable.

All in all, lava plains are just as widespread on Mars as are basaltic mare on the Moon's near side, and ocean floor basalts on planet Earth. Like these, martian lava plains are characterized by relatively flat and smooth expanses of overlapping flows, lacking large volcanic constructs, and showing little sign of their eruptive fissures – sealed and buried as they are by subsequent flows.

As is the case with the lunar maria, martian lava plains show subdued swells – wrinkle ridges – dozens of kilometers long and thousands of meters wide, resulting from compression and buckling of the expanses of lava. The swells on the martian plains are spaced 50 km apart on average. Viking 1 happened to land on the flank of such a ridge, a few kilometers from the crest: ground-based imagery shows a slope of a degree or two toward the northwest, where the horizon could well mark the crest of the ridge, or a structure subparallel to it.

Erosion on Mars

Based on crater counts, the Chryse and Utopia plains are old – probably in excess of 2 billion years – and erosion has transformed what were once smooth lava sheets into broken-up fields of rocks and boulders – some still sharp and angular, while others are sanded down by the abrasion of wind-blown dust. Patches of reworked 'mineral shavings' collect as pockets and dunes between the rocks, making up the well-mixed soil analyzed by the landing craft.

On Earth, lava flows break down to yield similar-looking rock fields, especially in the dry American Southwest, where we find remarkable analogs of martian landscapes. We know from these terrestrial sites that lava fields erode preferentially along cracks inherited from their cooling history, and which vary in spacing as a function of burial depth within the flow. At the surface, where cooling is rapid, cracks are closely spaced and break up the topmost part of a flow into a mosaic of small 'cobble stones'. Deeper into a flow, where cooling is slower, fractures are spaced farther apart. The size of loose boulders in a lava field therefore indicates their depth of origin – the larger, the deeper – or in other words how much of the flow was removed by erosion to unveil the boulders in question.

If we apply this model to the plains of Mars, we can guess their erosion history. On the Utopia landing site, the rocks are remarkably uniform in size – twenty to thirty centimeters – the kind of fracture spacing one would expect to find a meter or so below the surface. Apparently, the lava fields of Utopia were eroded by only a meter since their emplacement, and this gives us a good indication of the pace of erosion on the red planet. On Earth, the action of the sun, wind, and rain eats away the top meter of a flow in a couple of million years: this is the age of lava flows in Arizona, which best resemble the martian landscape, in overall appearance and rock size. But from crater counts, we know that martian plains are over 2 billion years old. In other words, erosion processes are one thousand times weaker on Mars than on Earth – which is why the flows look so pristine, despite their great age.

Martian cocktails

Trenches dug by the telescopic arm showed martian soil to be clay-like, and adhesive. Some mineral grains were seen to be magnetic: magnets mounted on the probe became covered with what appeared to be a hydrated form of iron oxide, of the

Table 6.1 *Composition of martian soil as measured by Viking 1 and 2, and compared with calculated analogs and reference mineral and rock on Earth. Percentages in mass*

	Mars (Viking)		Mars simulations		Earth	
	Chryse[a] (soil)	Utopia[b] (soil)	Analog[c] (calculated)	Analog[d] (mixed)	Nontronite[e] (Earth)	Diabase[f] (Earth)
SiO_2	44.0	43.0	43.6	44.6	45.6	48.2
Al_2O_3	7.3	7.0	6.9	5.5	5.1	9.4
Fe_2O_3	17.5	17.3	18.4	18.6	25.0	15.0
MgO	6.0	6.0	9.0	5.4	2.2	17.5
CaO	5.7	5.7	5.6	6.0	6.4	7.0
K_2O	0.0	0.0	0.0	0.1	0.1	0.4
TiO_2	0.6	0.5	0.9	0.7	0.5	0.8
SO_3	6.7	7.9	7.3	8.4	0.1	0.0
Cl	0.8	0.4	0.0	0.5	0.1	0.0
Total	88.6	89.8	91.7	89.8	85.1	98.3

The martian soils of columns 1 and 2 were sampled in 1976 by the Viking 1 and 2 automated landing craft, respectively on the sites of Chryse (22° N 48° W) and Utopia (44° N 226° W).

The samples were automatically analyzed on board the Vikings by fluorescence spectrometry. No data was recorded for sodium, phosphorus and manganese, which might account for as much as 3%. In total, 23 soil samples were analyzed by the Viking spacecraft on both sites: they turned out to be remarkably uniform.

[a] Chryse: average of two soil analyses, performed on the Chryse site, Mars, by Viking 1.

[b] Utopia: average of four soil analyses, performed on the Utopia site, Mars, by Viking 2.

[c] Calculated analog: theoretical mineral layout by computer, which would yield a chemical make-up similar to that observed by the Vikings. According to this model, the soil of Mars is mainly composed of ferromagnesian clays (47% nontronite; 17% montmorillonite; 15% saponite), a sulfate (13% kieserite), a carbonate (7% calcite) and a sprinkle of titanium oxide (1% rutile).

[d] Composed analog: mineral make-up of a sample obtained by physically mixing in the laboratory source minerals until chemical data ressembling the Viking figures were obtained. As in the calculated analog, ferromagnesian clays are dominant (49.5% nontronite ; 18.2% bentonite), with sulfate (13.7% kieserite); quartz (7.5%); calcite (4.9%); iron oxides (2.5% hematite and 2.4% magnetite); sodium salt (8% halite); and leucoxene (0.4%).

[e] Nontronite : terrestrial mineral which, taken separately, is the closest in composition to martian soil. As such, nontronite is the major ingredient of analogs (c) and (d). Nontronite is a ferromagnesian clay, and is a typical alteration mineral of lavas on Earth.

[f] Olivine diabase (New Jersey): altered dolerite (dolerite is an igneous rock of the basalt family). Among terrestrial igneous rocks, the olivine diabase is closest in composition to the martian soil.

maghemite family (γFe_2O_3). Rust-colored maghemite is a ferromagnesian clay, and such clays might make up as much as 75% of the martian soil.

We can further constrain mineral types and proportions by playing mineral 'scrabble' with the X-ray fluorescence data. The object of the game is to make 'words' with the measured alphabet of atoms, i.e. propose a mineral make-up that uses up all the atoms in the right proportions.

One way to play the game is to mix selected minerals in the laboratory, and adjust proportions until the mixture comes out with a chemical composition matching the Viking analyses. As shown in Table 6.1, the mineral 'cocktail' which yields the closest match to the Viking data is a mix of 47% *nontronite* clay (including its altered form *maghemite*); 17% *montmorillonite*, and 15% *saponite* clays; 13% of a sulfate named *kieserite*; 7% of *calcite* (calcium carbonate); and 1% of *rutile*, a titanium oxide.

Another approach is to run mineral equations in a computer, rather than physically mix powders, and likewise aim to match the element concentrations measured on Mars. By this method also, the ferromagnesian clays nontronite, montmorillonite, and saponite come out the clear winners.

Ferromagnesian clays are widespread on Earth. They typically form from the alteration of mafic rocks, and abound around tropical, basaltic volcanoes. Underwater lavas on Earth also break down to ferromagnesian clays: one particular form is *palagonite*, the altered basaltic glass of underwater eruptions.

On Mars, the ferromagnesian clays seem to be accompanied by kieserite and calcite. Kieserite, a hydrated sulfate of magnesium, would account for the high sulfur content of the martian soil. Calcite is a carbonate, that would store carbon dioxide, and lodge most of the weathered calcium. On Earth, sulfates and carbonates are found in evaporite basins – salt pans where ion-rich waters evaporate and precipitate their chemical loads. But they are also known to form in volcanic ground, where hydrothermal waters percolate ions to the surface, and precipitate a chemical residue in the topsoil. The rich concentration of kieserite and calcite in the martian soil could likewise be attributed to the percolation of ion-saturated fluids, in the waning stages of volcanic activity.

Meteorites from Mars

From the analysis of martian soil by the Vikings – and laboratory simulations on Earth – one gets a fair representation of the mineral make-up at the surface. But this assemblage of secondary, weathered minerals stops short of telling us what pristine, unweathered minerals made up the lava rocks in the first place.

This is where we get some unexpected help from heaven. Indeed, it appears that a handful of martian rocks have been conveniently delivered to our doorstep, after being blasted off the martian surface by some extraordinary event, crossing the Earth's orbit on a collision course, to eventually enter the atmosphere, and reach the ground as meteoritic debris.

Shooting stars from Mars? So far, close to a dozen meteorites have been classified as martian suspects: one is the *Chassigny* meteorite, which fell in France in 1852 (see

FIG. 6.4. A piece of Mars fallen to Earth? This unusual SNC meteorite that fell in 1852 near Chassigny, France, contains in its glass microscopic bubbles of gas that have much in common with the atmosphere of Mars, as measured by the Vikings. Other clues, including the meteorite's unusually young age of 1.3 billion years, point to a planetary origin. The rock is believed to have been blasted off the surface of Mars by an impact, before striking the Earth. The Chassigny meteorite is a pale olivine cumulate (dunite), with a thin coating of dark glass (in upper corner), inherited from the fiery reentry through the Earth's atmosphere. Credit: Courtesy of the Laboratoire de Minéralogie, Museum National d'Histoire Naturelle, Paris.

fig. 6.4). This is a light-colored, fine-grained meteorite made up almost exclusively of olivine – a rock type known as *dunite*, which typically forms in cumulate beds at the bottom of magma chambers.

Other meteorites of suspected martian origin include the *nakhlites* (the first sample of this class was found near the town of Nakhla in Egypt), which are cumulates of pyroxene crystals – another common concentration that occurs in magma chambers; and the *shergottites*, (named after the town of Shergotty in India), which are basically basalts (assemblages of olivine, pyroxene, and feldspar). A couple of meteorites found in Antarctica round the list, including a pyroxene-rich specimen identified in 1994, which contains traces of carbonate – thought to result from hydrothermal activity in the zone of genesis.

Shergottites, nakhlites, the Chassigny object, and the Antarctic meteorites are grouped under the denomination *SNC*, and share a variety of characteristics that point to a martian origin.

Firstly, their mineral assemblages point to a magmatic genesis at high pressure – a condition that occurs only at depth within sizeable planets, not inside small asteroids which are the source of most other meteorites. The cumulate texture of the SNCs also requires a significant gravity field to separate crystals of differing densities, again pointing to a massive planetary parent. Therefore, large objects like the Moon, Mars, Mercury, and Venus head the list of suspects for the genesis of the SNC rocks.

Secondly, SNC specimens contain small bubbles of trapped gas, which – when punctured and analyzed in the lab – yield atomic and isotopic ratios that match those of the martian atmosphere, as recorded by the Vikings.

Lastly, SNC meteorites all share the same young age of 1.3 billion years, a figure which matches our estimates for the age of the martian plains, and rules out the much older surfaces of the asteroids and the Moon as possible sources for the objects.

The only catch is to explain how chunks of igneous rock were blasted off the surface of Mars at speeds in excess of 5 km/s, pulled away from the planet's gravitational field, and ultimately reached the Earth. Volcanic eruptions as a propulsion mode can be ruled out, for sizeable chunks of lava cannot be accelerated by the expanding gases of a martian eruption at speeds any greater than a few hundred meters per second. Large meteoritic impacts on Mars would provide a much better propulsion mechanism, smashing the surface and sending debris flying outwards on a cushion of expanding volatiles. Speeds in excess of 5 km/s are not unreasonable in the impact scenario, and deep-digging impacts would also explain how lower crustal rocks – like dunite and pyroxenite – were included in the ejecta.

In the heart of the mantle

If SNC meteorites are truly rocks from Mars – and we have every reason to suspect so – then they can tell us a great deal about the primary mineral composition of the red planet, from the crust down.

Guessing the composition of a planet's deep mantle from surface rocks is a familiar exercise for mineralogists, since lavas are derived from the underlying mantle according to well-known chemical and physical laws – the rules of partial melting.

On Earth, the exercise is well constrained because we can check our answers by studying the behavior of seismic waves crossing the mantle, which gives us an independent estimate of its mineral composition. We are also fortunate on Earth to have direct access to some mantle rocks at the surface – small samples coughed up by deep-rooted eruptions (unmolten nuggets called *xenoliths*); as well as entire rock layers from the upper mantle, thrust up over the crust in tectonic compression zones (*ophiolite series*).

On Mars we are not so lucky as to have direct access to the mantle, but we can rely on chemical and physical equations to write out its estimated composition, plugging in the mineral data from SNC meteorites, as well as reference data from 'ordinary' meteorites – the building blocks from which all planets are made. Add to this a few clues from the Viking soil analyses (high iron and sulfur content), and we can come up with a reasonable guess for the composition of the martian mantle: 50% olivine; 25% orthopyroxene; 15% calcium-rich clinopyroxene; and 10% garnet. This would

FIG. 6.5. Ceraunius Tholus (24° N, 97° W) is a 100 km shield volcano on the periphery of the Tharsis plateau (for location see map, fig. 5.3). Like its two neighbors Uranius Tholus and Uranius Patera, Ceraunius Tholus is about 3 billion years old and precedes the build-up of the giant Tharsis shields, more central to the plateau. Ceraunius has a 20 km caldera, with a lava channel descending to a basin at the foot of the shield – perhaps a reworked impact crater. Another impact crater is visible south of the volcano, with a fluid ejecta blanket. Credit: NASA/JPL, courtesy of Brown University.

be similar to the mantle composition of the Earth, with slightly more iron and less magnesium.

Of course, these are educated guesses at best. Because we ignore the moment of inertia of Mars – a fundamental physical parameter – we have little way of telling the distribution of masses inside the planet, and hence the density of its mantle. Mars could well have more iron in its mantle than the Earth does, but it is not proven. Landing on Mars and analyzing suites of rocks will help solve the issue.

Melting scenarios

In the meantime, we are left to play with hypotheses and models. Starting with the above estimated composition for the martian mantle, we can play out different melting scenarios, and predict what range of lavas we would expect to find at the surface.

At the onset of partial melting, the martian mantle will yield a magma rich in magnesium and iron, with a substantial concentration of sodium and potassium – the familiar make-up of *alkaline basalts*. As we raise the proportion of mantle that enters the melt (by raising the temperature, for instance), the magma gets richer in iron, and the result is high temperature, very fluid komatiite lavas.

Nothing very exotic so far: on Earth as well we get alkaline basalts for small percentages of partial melting, and less alkaline lavas as we melt increasingly larger proportions of the mantle. When the proportion of melting reaches 40%, one also gets komatiite lavas on Earth.

A difference does exist between the terrestrial and martian trends, however, concerning the degree of ease with which fluid komatiite lavas might be generated.

VOLATILE CONTENT OF MARTIAN MAGMA

On Earth, melting 40% of the mantle to get komatiite lavas is a difficult exercise, because very high temperatures are needed (1450 °C to 1700 °C) to achieve such high melting ratios. Only in the first half of the Earth's history, when the mantle was arguably hundreds of degrees hotter than today's, was komatiite lava produced – we find komatiite outcrops in precambrian terrain, as on the Canadian shield where they are 2 billion years old.

On Mars, komatiite could be much more widespread: if the red planet's mantle is indeed much richer in iron, this type of magma would be produced at lower melting ratios than on Earth – i.e. at lower mantle temperatures. Komatiites might then be the dominant lava type we see at the surface, rather than its close cousin basalt.

The difference between komatiite and basalt is slim: komatiite is more fluid than basalt and flows over longer distances (other conditions like discharge rate and temperature being equal). This is in keeping with the great run-out distances of martian flows, as observed on Mariner and Viking orbital imagery.

A second characteristic of komatiite is that its magma can carry substantial amounts of sulfur in solution – a few tenths of a percent – because it is iron rich, and sulfur and iron tend to 'stick together' (sulfur is a siderophile). During komatiite emplacement, sulfur will ultimately separate from the cooling lava to form its own pockets and veins. The process is well known on Earth, where sulfide pockets are found lodged under komatiite flows, as ore sills of pyrrhotite and pendlantite. Sulfur veins are particularly sensitive to dissolution and remobilization during hydrothermal activity, and sulfur-rich solutions end up percolating to the surface. This might explain the high sulfur content of the martian soil, as measured by the Viking landers.

Volatile content of martian magma

By piecing together the Viking chemical data and the mineralogical clues from the SNC meteorites, we can then get a fair idea of the nature of martian lavas. One important piece of information is still missing to define martian volcanism, however: the nature and quantity of the volatiles that flush the melts, and drive eruptions at the surface.

Were the eruptions gasless and peaceful, with ribbons of lava pouring out of fissures undisturbed? Or else were there lava fountains on Mars, driven by escaping gases, with clouds of ash climbing to great heights through the atmosphere, and *nuées ardentes* rolling down the slopes of the giant volcanoes?

We have seen that magma behavior is conditioned both by volatile content, and by the confining pressure. At depth the volatiles remain in solution, but as the magma ascends to the surface, the drop in pressure causes the volatiles to come out of solution and nucleate as bubbles. This is the 'champagne syndrome': as the pressure keeps dropping, the bubbles expand until they turn the magma into a foaming emulsion, and blow it skyward. For a typical basaltic magma on Earth, calculations show that explosive disruption will occur at the surface for water contents (by mass) of 0.06% and higher. Using heavier carbon dioxide as a driving gas, the minimum content that leads to disruption is a slightly higher 0.1%.

FIG. 6.6. Close-up of martian lava on Viking 2's landing site (Utopia Planitia). A dark sandy trough crosses the frame in the foreground, part of a local fracture pattern affecting the topsoil of the site. Behind the sand pocket, lava blocks are prominent and highly vesicular. The central rock is one meter long and appears banded – its left extremity darker and more vesicular than the rest of the rock. Perhaps this extremity represents the top of the lava flow, that underwent more gas exsolution and frothing than the rest. Credit: NASA/JPL, courtesy of Brown University.

On Mars there is a major difference in that atmospheric pressure at the surface is 200 times weaker than on Earth – 6 hectopascals in the martian plains, even less at the top of volcanoes. Unconstrained by such low pressures, magma gases expand considerably, so that lesser amounts of volatiles are needed on Mars to foster magma disruption at the surface – six times less, in fact, than on Earth. Hence, a mere 0.01% of dissolved water in martian magma will lead to disruptive behavior, versus 0.06% on Earth.

It seems likely, therefore, that explosive volcanism was commonplace on Mars, as long as there were 'reasonable' amounts of volatiles in the mantle. This seems to have been the case, since SNC meteorites from Mars contain hydrated minerals (amphiboles in the case of Chassigny), which is good proof of a water-bearing mantle. There is also ample evidence of water near the surface of Mars, trapped as permafrost. Carbon dioxide is also omnipresent, frozen in the polar ice caps, and as the atmosphere's main gas. This abundance of water and carbon dioxide on Mars points to millions of years of volcanic degassing, and to a volatile-rich mantle.

There is another line of evidence that points to volatiles: the spongy aspect of lava rocks, as photographed on the surface. Many of the lavas viewed by the Viking landers are vesicular in texture – the mark of gas exsolution during the formation of the rock (see fig. 6.6). There are other mechanisms to explain the vesicles, such as differential plucking of mineral grains by wind-driven blasting, or chipping of the rocks by cycles of freezing and thawing. But bubbling gas is considered the simplest, best explanation for the pitted appearance of the martian lavas, and this reinforces the interpretation that volatile-rich, explosive volcanism was commonplace on the red planet – at least through certain epochs of its history.

We would expect degassing to have ruled the early days of martian volcanism, as volatiles were progressively transfered from the deep mantle to the upper crust and atmosphere. As the mantle got rid of its volatiles, volcanism would have quieted down over time from gas-rich and explosive, to gas-poor and effusive. This is at least the expected trend. In detail, magma interaction with volatiles was probably much more complex. There certainly were cases, for instance, when otherwise gas-poor magmas intersected volatile-rich sediments on their way to the surface.

On Earth, this occurs when ascending magma crosses water-soaked ground: the sudden expansion of steam can turn the eruption into an explosion, blasting a shallow crater at the surface, known as a maar. On the red planet, we would expect similar behavior every time magma crossed the permafrost layer, especially at high latitudes where the layer is known to be thickest. It is thus fair to say that eruptions on Mars were extremely varied, both in time and space.

A history of martian volcanism

We can attempt to trace the evolution of martian volcanism through time by studying the changing aspect of its volcanoes, from oldest to youngest.

One finds the oldest volcanoes of Mars both in the equatorial and northern provinces of Tharsis and Elysium – where they are half buried by younger flows – as well as in the cratered highlands of the southern hemisphere, where most cluster around the Hellas impact basin.

Hellas is one of the large basins that date back to the shattering of the early crust by large asteroids – the heavy bombardment phase that swept through the Solar System 4 billion years ago. Crustal blasting by large impacts was one way to create the right kind of fracture zones for magma to reach the surface. This was certainly the case around the Hellas basin, where lie four of the landmark volcanoes of the period – Peneus, Amphitrites, Hadriaca, and Tyrrhena Paterae. All took advantage of the fracture rings developed by the giant impact.

These antique Hellas volcanoes are broad, low-incline shields with wide, shallow calderas; and intriguing 'dandelion' channels gouging the flanks. Hadriaca (see fig. 6.7) and Tyrrhena (see fig. 5.7) are the most representative: both are shields 200 to 300 km in diameter, centered around large, shallow calderas. The flanks of these vast aprons are only fractions of a degree in slope, and are gouged by wide channels that betray the volcanoes' past environment. Indeed, these are not lava channels – which on Mars are typically one or two kilometers wide, and fan out into lava fields – but corridors much larger in size (up to 10 km), and separated by broad ridges.

The channels at Hadriaca Patera extend predominantly to the southwest, influenced by the regional slope toward the Hellas basin. Tyrrhena Patera, closer to the basin, has its channels radiating in all directions, which earned it its nickname 'Dandelion'. The wide channels on these aprons imply water erosion through soft, easily gouged material, which supports two intertwined theories: that Mars

FIG. 6.7. Hadriaca Patera (31° S, 267° W) is a 300 km-wide volcanic apron, centered around a subdued, central caldera (upper right margin). Impact craters pepper the slopes, bracketing the age of the edifice around 3 billion years. Like other highland patera, Hadriaca is characterized by wide channels separated by ridges, probably formed by water-sapping of soft ash strata. Pyroclastic patera of the sort erupted at a time when volatiles were abundant in the martian magma or in the assimilated crust. Notice the eroded basins of the Hellas plateau, left and bottom. Credit: NASA/JPL, courtesy of Brown University.

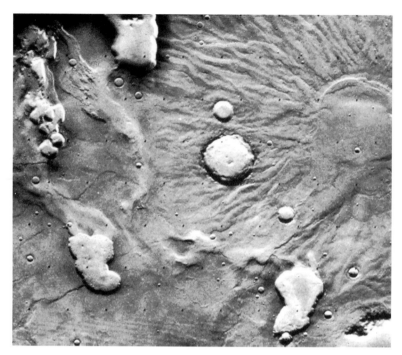

experienced warmer environments in the past (the Noachian period), when higher atmospheric pressures allowed liquid water to flow – at least underground; and that early volcanoes were built of soft ash, a consequence of their volatile-rich, explosive activity.

On Earth, large volcanoes with low relief and soft, erodable flanks make up the class of pyroclastic, explosive features known as resurgent calderas (see Chapter 2). Most of these aprons of ash and welded tuff exhibit well-developed patterns of gullies and rivers, the best examples being the resurgent calderas of the American Southwest; Mexico; the Andes (see fig. 2.10); New Guinea; and New Zealand. Hadriaca, Tyrrhena Patera, and other early martian volcanoes were probably blow-out calderas of this type: eruptions on Mars took the form of massive ignimbrite flows – clouds of pulverized hot rock flowing far and wide down the slopes.

On Earth, explosive behavior is associated almost exclusively with siliceous, viscous magmas that can store sizeable amounts of volatiles in solution. Ash blow-outs are therefore rhyolitic or dacitic in composition. It is much rarer to observe pyroclastic flows of basaltic composition on Earth, because of the limited ability of basaltic magma to soak up enough volatiles at depth to trigger violent disruption at the surface – one exception being the basaltic ignimbrite of Crater Lake caldera in Oregon.

The situation is different on Mars, where eruptions take place in near vacuum: even the low volatile content of an average basaltic magma is enough to disrupt the batch explosively at the surface, and drive columns of ash, nuées ardentes, and other ignimbrite flows.

We can view this first epoch of martian volcanism, 4 to 3 billion years ago, as a showcase of explosive eruptions, ash flows spreading around calderas, and releasing large amounts of volatiles into the atmosphere. The soft, porous tuffs of the patera could well have acted as reservoirs for these dumped volatiles – mostly water and carbon dioxide, crystallizing as permafrost ices in the surface layers. Under favorable conditions, one can imagine that the water crystals mixed with ash were piled to such depths that the load pressure turned the ice to liquid – especially under the high heat flows typical of volcanic areas. The channels of Tyrrhena and Hadriaca might then be the result of large-scale surface sapping, during periods of massive water runoff.

Perhaps these volatile reservoirs, permeating volcanic ash beds, guaranteed sustained longevity to explosive volcanism by providing sources of water vapor and carbon dioxide to otherwise gas-poor, ascending magmas. In essence, volatiles would have been largely recycled in the martian crust, contributing multiple episodes of explosive activity.

Plate tectonics on Mars

Explosive eruptions were certainly pervasive in the early days of martian volcanism, but there were many calmer, effusive eruptions, that show up as extensive lava fields in the northern latitudes. Researchers even consider the possibility that basaltic ridge activity – comparable to ocean-floor spreading on Earth – took place at that time.

If plate tectonics had any chance of developing on the red planet, it would have been in the early era of high heat flow. Because Mars is so much smaller than Earth, it could only sustain a high heat flow and vigorous mantle convection for a short period of time, before the planet's heat effectively bled out, and the cooling planet switched to more restrained forms of volcanism.

If early rifting and lava flooding did occur on Mars, activity was apparently focused in the northern hemisphere, where large, lava-filled basins are found. Perhaps most large impacts occured in the northern latitudes, creating extra heat there, and faulting the crust, whereas the southern hemisphere remained thicker-crusted, and hostile to magma ascent (except at local impact basins like Hellas).

One can even imagine the onset of large-scale crustal rifting in the northern basins, because of the thinning of the lithosphere. The vastness and flatness of the northern provinces do remind us of our own ocean basins on Earth, as do their step-faulted margins, where they rise to meet the cratered plateaux to the south – a staircase pattern reminiscent of rift margins on Earth.

Taking the sea-floor spreading comparison one step further, scientists have pointed out an unusual topographic rise poking out of the lava fields, halfway between Tharsis and Elysium: the north–south ridge of Phlegra Montes (195° W, 30–46° N).

Centrally positioned in the middle of a major basin, Phlegra Montes might be an extinct transform fault, i.e. the expression of a past episode of ridge-like, extensional volcanism. One might even speculate that the three Tharsis volcanoes of Arsia, Pavonis, and Ascraeus Mons fall on a line because they stand behind an ancient subduction plane, operative in the early days of Tharsis.

This speculative plate tectonic model (see fig. 6.8) calls for the splitting of the martian crust into two plates in Noachian times, with its rifting margin running along the Memnonia, Cimmeria, and Tyrrhena provinces. The opposite side of the plate would fall about even with the three Tharsis volcanoes (Ascraeus, Arsia, and Pavonis Mons), defined in this model as a subduction zone or island arc. Another subduction zone might have operated under the Arabian plateau, accounting for some of its ridged plains.

In this reconstruction, older highland crust was subducted, as younger, thinner crust was created to replace it, with Gordii Dorsum a ridge remnant from this ancient spreading phase. Finally, when plate motion and subduction shut down – allegedly when most of the ridge was swallowed up by the trench, as frequently happens on Earth – two ridge segments stayed active in the middle of the basin, building up Alba Patera and Olympus Mons.

Although this plate tectonic framework of early martian volcanism is highly speculative, the importance of large-scale tectonics should not be neglected on Mars, and the vast graben of Valles Marineris is there to remind us that extensional forces have affected the martian crust into much more recent times.

From basins to shields

The first epoch of martian volcanism can then be summarized as one of explosive ash flows and effusive lava fields – perhaps stopping short of disrupting the crust into plates, but certainly creating some ambiguous morphologies in the northern basins, as we just saw. The emplacement of lava flows in the martian plains is assumed to have peaked in the upper Noachian and early Hesperian age, some 4 to 3.5 billion years ago, although the record is poorly preserved because of the overprint of impact craters. From recognizable lava units, rates of volcanic resurfacing seem to have averaged one square kilometer of lava per year – only three to four times lower than volcanic rates on Earth today.

After this initial, sustained activity, Mars felt the steady decline of its heat flow, and the volcanic regime switched from basin-wide to much more centralized activity. The Elysium and Tharsis uplifts might have developed as a result of this shift of mantle convection to a few rare but large plumes.

The first volcanoes to grow over these probable hot spots, about 3 billion years ago, were Hecates Tholus, followed by Elysium Mons and Albor Tholus on the Elysium uplift; and in the Tharsis region Ceraunius (see fig. 6.5), Uranius, and Tharsis Tholus, followed by Biblis and Ulysses Paterae. Also taking place at the time was the phenomenal creation of Alba Patera, building up its shield flow after flow over hundreds of millions of years.

As we pointed out in the previous chapter, the great size and longevity of martian volcanoes is due to their static growth over feeding hot spots, where they stay connected to their magma source, rather than moving away from their hot spot sources atop shifting tectonic plates, as they do on Earth. On Mars, this long-term coupling with magma sources at depth allowed volcanoes to build up over

FIG. 6.8. A speculative reconstruction of plate tectonics in early martian history, based on the geometry of the highland/lowland dichotomy and structures in the volcanic lowlands. In the left sketch, this two-plate model shows a boreal plate (stippled) separated from the rest of the crust by spreading ridges along the eastern dichotomy (top) and subduction zones along the Tharsis plateau and the Arabia highlands. In the right sketch, the boreal plate has split into two halves which continue to subduct, faster than crustal creation at the ridges. Olympus Mons and Alba Patera would start their growth on ridge segments of the shrinking, subducting plate. Credit: Courtesy of Dr Norman H. Sleep, from N. H. Sleep, 1994, Martian plate tectonics, *J. Geophys. Res.*, **99**, 5639–55.

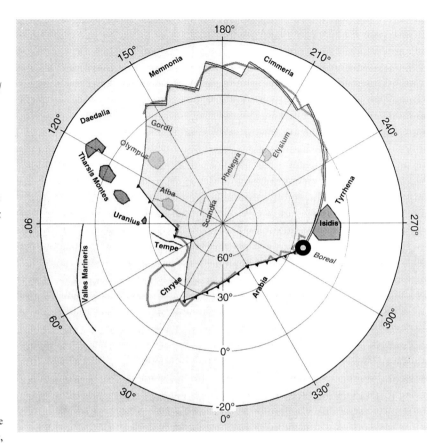

considerable periods of time, as long as the mantle plumes were active – hundreds of millions of years, as opposed to hundreds of thousands of years for shifting terrestrial volcanoes.

What were the eruptive styles of these Hesperian-age volcanoes on Mars? We can imagine that shield-building was accompanied by vigorous lava fountaining at the vents, and even more violent explosive activity, since volatiles were certainly not lacking at the time. Signs of pyroclastic activity might be recorded at Apollinaris Patera, a low shield south of Elysium, 200 km across with a 70 km-wide caldera. Apollinaris displays a fan-shaped array of ridges and troughs to the south: these could well be immature fluvial channels, or volcanic density currents – both characteristic of a volatile-rich environment.

Hecates Tholus, on the Elysium uplift, also displays aprons of ash-looking material down its slopes and around its base. And in the Tharsis region, the oldest patches of Alba Patera also seem to be soft strata deposited by pyroclastic activity. In fact, long-lasting Alba spans the interval of time when explosive, volatile-rich eruptions gave way to degassed and more peaceful lava flow regimes.

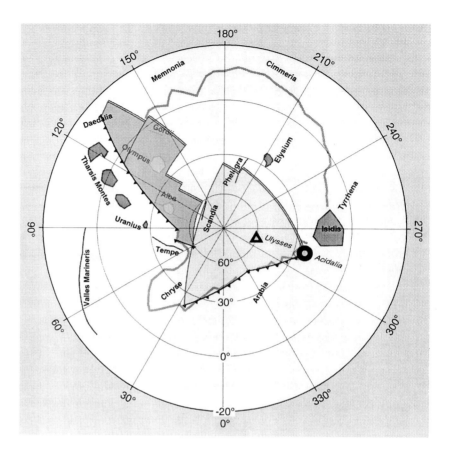

Alba the great

The oldest layers of Alba Patera crop out on the lower flanks, show an ash-like consistency, and are apparently marked by water sapping. These strata stand out as a broad ring extending 250 to 450 km from the central caldera. One explanation for their origin is that they are air-fall deposits, from violent pyroclastic blow-outs at the caldera.

We can imagine the sight that Alba Patera offered at the time. Calculations indicate that a cloud of ash would climb five times higher on Mars than on Earth (because of the lower gravity field, and especially the lower atmospheric pressure), as well as spread over five times the area. At Alba, columns of fine ash were probably pumped up to altitudes of 100 to 200 km, as opposed to 30 to 40 km in terrestrial eruptions.

But even if we accept great cloud heights on Mars, some of the ash fields at Alba spread so far down wind from the central caldera that other mechanisms are sought to explain their exceptional range. In fact, ash clouds on Mars should collapse readily under their own weight, poorly sustained as they are by the low atmospheric pressure. Such collapses would send base surges rolling over great distances, which would explain the extensive ash fields.

FIG. 6.9. Tharsis Tholus is a steep volcanic shield on the eastern border of the Tharsis plateau (13° N, 91° W, see location on fig. 5.3). It is 110 × 170 km in size and has a large summit caldera, central to a north-south graben cutting through the edifice. The lower slopes are truncated by more recent lava flows. Notice the patch of darker ground east (to the right) of the volcano – probably a wind streak in its lee. Tharsis Tholus was one of the last volcanoes formed on Mars before the four giant shields of Tharsis grew to the west, and monopolized the volcanic activity. Credit: NASA, courtesy of the National Space Science Data Center, through the World Data Center-A for Rockets and Satellites (Viking Experiment Team Leader: Dr Michael H. Carr).

On Earth, pyroclastic flows can cover distances up to 150 km in the best of cases. On Mars, the nuées ardentes at Alba could have flowed 400 km according to calculations. Indeed, in a ground-hugging ash flow, the kinetic energy is proportional to the square power of the gas velocity at the vent. On Mars, gas velocities are 1.5 times higher than on Earth, due to the increased expansion under lower atmospheric pressure. Elevated to the square power, this gas velocity factor means that pyroclastic flows on Mars can travel over twice as far as on Earth, reaching run-out distances in excess of 300 km.

After the ash flows of their early, volatile-rich eruptions, the hot spot volcanoes of Tharsis and Elysium switched to more effusive activity, and built up their shields with lava flows.

Over the last billion years, activity was restricted to only four sites on the Tharsis plateau, building the giant shields of Arsia, Pavonis, Ascraeus, and Olympus Mons. These last volcanoes to grow on Mars apparently did so in a staggered fashion, if we are to believe the age pattern of impact craters on their shields: activity at the southern Arsia Mons ceased 700 million years ago; whereas Pavonis eruptions lasted until 300 million years ago; and those of Ascraeus until 100 million years; not to mention Olympus Mons, where the youngest summit flows might be as young as 30 million years. The age progression seems to underscore a geographical trend through time, from oldest activity in the southwest to youngest in the northeast. Could this be a mere coincidence in the age line-up, or the surface trace of a real progression of a hot spot plume under the Tharsis uplift, with the magma source migrating southwest to northeast?

Another intriguing observation is the near-equal elevations reached by all four volcanoes, although they are hundreds of kilometers apart: Arsia, Pavonis, Ascraeus, and Olympus Mons all peak at 27 000 m above the reference plains. Rather than a coincidence, this might well indicate that all four pump their magma from the same depth in the mantle – the maximum height of a volcano being directly proportional to the depth of the magma source, and to the difference in density between the ascending buoyant magma, and the encasing rock. If we set the density difference equal to 10% (an average density difference between molten and solid silicates), then a volcano altitude of 27 000 m equates to a magma source depth of 250 km. We can speculate that the giants of Tharsis progressively grew taller as the planet slowly cooled and the melt zone migrated to the 250 km level, where the last dregs of magma were pressed.

We will have to wait until new probes reach Mars before we can check up on these many theories, and draw a more accurate picture of planet Mars. At the time of this writing, the Global Surveyor and Pathfinder craft are scheduled to reach Mars in the summer of 1997. The latter is a landing craft with an automated rover, targeted to explore the ancient, outwash channels of Ares Vallis. There are also plans for a Russian probe in 1998, with both a larger rover and an automated balloon, that will drift from place to place across the landscape, take photos and make measurements. Truly, the exploration of Mars has only begun.

7 Venus unveiled

FIG. 7.1. Mosaic of
the surface of Venus,
Magellan radar imagery.
Credit NASA/JPL.

Venus, our sister planet

Venus is the closest planet to Earth: our neighbor circles the Sun on an orbit averaging 108 million kilometers in radius (versus 150 million kilometers for the Earth), an orbit which it covers in 225 days – close to seven and a half months.

Faster-moving than our planet – on a smaller ellipse around the Sun – Venus laps us every nineteen months. During the months when it is chasing us, Venus is visible in our skies as a bright evening star, before disappearing in the Sun's halo. After overtaking us, Venus emerges from the Sun's glare ahead of us, as an early morning star. Also known as the Shepherd's star, Venus is the brightest object in the sky after the Sun and the Moon, on those months when it is favorably lit.

Venus is not only our bright neighbor. It is also the world that resembles us most in terms of size: its diameter falls short of the Earth's by only 652 km – 12 104 km versus 12 756 km. This might seem like a negligible difference on a linear scale, but it takes on added dimensions when one reasons in terms of areas and volumes: the area of Venus amounts to only 90% of Earth's, and in volume the ratio falls to 80% (see

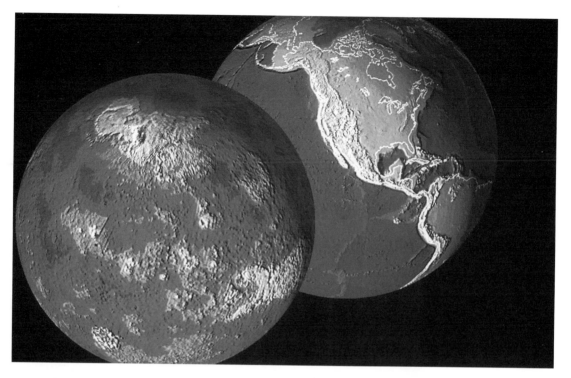

FIG. 7.2. Comparison of size and topography of Venus and Earth. The diameter of Venus is slightly inferior to that of our planet (12 104 km versus 12 756 km), which translates into a volume 80% that of the Earth. The opaque cloud veil is lifted in this artist's rendition, revealing the main topographic provinces of our sister planet. The highest elevations are lightest, midlands in mid-tones (volcanic rises and plateaux), and lower than average elevations darkest (plains, basins and troughs). The highland 'cap' in the northern hemisphere is Ishtar Terra, with the Maxwell Montes range in white. Along the equator is the highland chain of Aphrodite Terra. Credit: NASA/JPL, courtesy of Brown University.

fig. 7.2). Hence, Venus is more of a 'little sister' than a twin, with a personality quite its own.

For one, Venus is not the pleasant, tropical paradise it was made out to be by the poets, who were lured by its 'warm' proximity to the Sun, and its mysterious veil of clouds. Quite the opposite, Venus turned out to be a most hostile world, crushed by a hellish atmosphere 80 times the weight of the Earth's (8 MPa), and heated by the greenhouse effect to over 700 K (430 °C). The greenhouse heating is principally the doing of carbon dioxide (the atmosphere's main constituent), with the added contribution of sulfur dioxide and traces of water vapor – all of which let solar light pass through to the ground, but trap the outgoing infrared radiation, and drive the temperature up. Day or night, the temperature never falls below 700 K at the surface of Venus, and occasionally reaches 750 K (480 °C).

As for the high pressures, they are comparable to those encountered by our deep ocean submarines: with a 5 MPa pressure atop its highest mountains, 8 MPa in the plains, and 10 MPa in its deepest basins, the atmosphere of Venus exerts crushing forces comparable to water pressures in our oceans at depths of 500 to 1000 m.

Descent into hell

Venus is veiled by a global cloud cover, which conceals the entire surface from geologists. Spectral analyses of solar light reflected off the cloud tops inform us that the ceiling of the opaque layer is at an altitude of 60 km – where pressure already reaches one atmosphere (0.1 MPa) – and is composed of fine droplets of sulfuric acid, hydrochloric acid, and hydrofluoric acid in suspension. Not exactly the tropical paradise chanted by the poets!

The base of the cloud layer ends 30 km from the ground (this we will learn from plummeting space probes), and the lower atmosphere of carbon dioxide is then crystal-clear to the ground.

The first space probes sent to Venus by the Soviets were built with the sturdiness of deep-diving submarines, to withstand the inhospitable pressures and temperatures. The electronics of those early probes were so aggressed by the adverse conditions that Venera 1 through 4 broke down during the landing sequence, and ceased to emit before touchdown.

It was not until Venera 7, in 1970, that a probe from Earth finally survived to the surface. Venera 7 landed on the night side of Venus, and before its instruments failed, radioed back data for 22 minutes, including the first weather report from another planet: 750 K (480 °C); and 9 MPa (90 atmospheres) of pressure.

Two years after this initial success, Venera 8 doubled the score, landing in the same region as its predecessor – the plains of Lavinia, at the foot of the Alpha plateau – but this time in daylight. The second broadcast from the surface of Venus lasted close to an hour, with an automatic, chemical analysis of the ground, and a new weather report: 740 K and 9.3 MPa.

The analyses of the atmosphere performed by the probes indicated a composition dominated by carbon dioxide (96%); with substantial nitrogen (3.5%); and traces of oxygen (0.01%), sulfur dioxide, water vapor, and argon. The lower, heavy atmosphere appeared driven by very sluggish winds blowing at 1 m/s (3 km/hr), and the luminosity at the surface – measured by a photometer – was so reduced by the cloud cover that the Soviets compared it to the light of an overcast, winter day in Moscow. To compensate, floodlights were added to the design of future probes, for photographic purposes.

Landfall on a volcano

The next generation of Soviet probes – Venera 9 and 10 – gathered the first black and white images of the surface of Venus. Venera 9 landed on the northeastern slope of the Beta Regio uplift (31° N, 291° W). Three days later, it was joined by Venera 10 (16° N, 291° W), 2000 km to the south (see map, fig. 7.4). Both sites were volcanic plains – as later revealed by orbital radar mapping – on the edge of Beta's giant volcanic shield.

Each probe radioed back two panoramic images of its landing site. Venera 9, perched on a 30° slope, revealed a rock-strewn landscape with flat-looking rocks a

FIG. 7.3. First image of the ground of Venus, collected by the Venera 9 Soviet probe on October 22, 1975. The probe landed on a slight incline, at the foot of the Beta Regio volcanic rise (see fig. 7.4 for location). The 174° panorama displays a chaotic array of lava blocks, believed to be a talus deposit. Credit: USSR Academy of Science, courtesy of the Vernadsky Institute and Brown University.

few decimeters across, resting on a talus of cobbles and soil (see fig. 7.3). A recent interpretation is that the probe landed on the slope of one of the many faults that criss-cross the area (as later revealed by Magellan imagery), faults that allegedly broke up the flows of ancient volcanic plains.

Venera 10, to the south, scanned a mosaic of layered rock 'tiles', looking again very much like the broken-up surface of a thin lava flow, with pockets of soil nested between the outcrops. Later radar imagery from orbit showed the landing area to be quite volcanic indeed (see fig. 7.4).

Completing the picture, the two probes performed radiogenic analyses of the ground, by measuring the gamma rays emitted by the surface. The concentration of radioactive elements was found to be very close to that of terrestrial basalts: 0.9% potassium; 0.00005% (0.5 ppm) of uranium; and 0.0004% (4 ppm) of thorium on the Venera 9 site; and a little less potassium (0.3%), but the same low values of uranium and thorium on the Venera 10 site. On Earth this chemical range is typical of rift-valley basalts: the moderate alkaline content, in particular, is reminiscent of lavas in Africa's Afar rift.

Building on their early successes, the Soviets launched a new salvo of Venus probes in 1978 (Venera 11 and 12), this time joined by two American probes – the Pioneer Venus mission. Of the two American spacecraft, one was to stay in orbit around the veiled planet and map its altimetry by radar, while the other was to dive into the atmosphere and radio back chemical and physical data.

On the Russian side, Venera 11 and 12 failed to collect new pictures of the ground – a faulty circuit deactivated the cameras on both probes – but on their way down to the surface the landers made an intriguing discovery: as they broke through the cloud deck – at an altitude of 30 km – and down to a couple of kilometers from the surface, the electromagnetic sensors on board the probes detected an uninterrupted stream of distant lightning bursts – up to 25 events per second. Soon after touchdown, Venera 12 also registered a distant, yet powerful thunder clap which rolled over the site for many seconds.

This confirmed earlier observations by Venera 9, which had detected what seemed at the time like lightning flashes illuminating the dark side of the planet, during the probe's final approach.

These electromagnetic storms on Venus were puzzling to the scientists, since the cloudy planet lacked large rain drops – best known to trigger electrical discharges of this sort. Without large drops, it became difficult to explain the electrostatic

FIG. 7.4. Map of the
Venera landing sites. The
first ground shots of
Venus were taken in the
eastern foothills of Beta
Regio (see fig. 7.3); the
next four probes landed in
the equatorial plains east
of Phoebe Regio (see fig.
7.5). Credit: Reproduced
with permission from
Planetary Volcanism by
Peter Cattermole,
published in 1989 by Ellis
Horwood, Chichester.

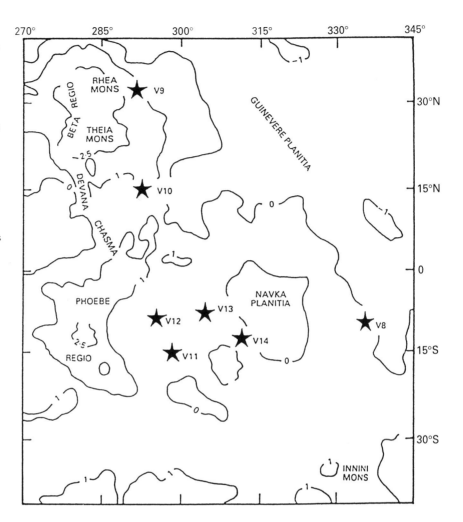

build-up. . . unless one resorted to large ash flakes, of the kind produced in plinian
volcanic eruptions!

Venus drops her veil

Volcanic eruptions on Venus, with thick clouds of ash climbing through the
atmosphere, laced with lightning? An exciting prospect for geologists, who looked
toward Pioneer Venus for confirmation.

The American probe became the first spacecraft to achieve Venus orbit, and
began its topographic mission soon after the Venera landings, bouncing radar waves
off the planet's surface to measure altitudes – and thus relief – with a vertical
accuracy of 100 m. The horizontal resolution was crude – each measurement
covering an area 100 km by 100 km wide – but sufficient to map out the broad
topographic provinces of our sister planet.

Two areas of relatively high ground could be made out: Ishtar Terra in the northern latitudes, the size of Australia; and Aphrodite Terra, a wide belt of swirling ridges wrapping one third of the way around the globe (15 000 km), inclined 25° on the equator, and half the size of Africa (see fig. 7.2).

These plateau-like swells rose an average of 3000 to 4000 m above the mean reference level, bordered in a few limited areas by trench-like troughs, plunging to 4000 m below the reference level.

A third, more restricted region of high ground was found well apart from Ishtar and Aphrodite. Named Beta Regio, it was resolved into a rather domical uplift, capped by two volcano-looking shields, about 300 km in diameter, and rounding off 4000 m above the plains: Rhea and Theia Mons. The region appeared divided by large rifts, notably one deep chasma running along the western flank of Theia Mons. More detailed mapping by Earth-based radiotelescopes, in the years immediately following the Pioneer mission, showed the Beta shields to be dimpled by summit depressions, tens of kilometers wide; as well as scoured by radial markings down the flanks, and extending into the graben. Summit calderas and lava flows?

Theia and Rhea Mons became the first two candidate volcanoes on Venus. The exciting hypothesis was reinforced by the lava-looking rocks photographed by Venera 9 and 10, which landed in the plains due east of the massif. And to top off the argumentation, Beta Regio was proposed as the source area of the lightning noise recorded by Pioneer Venus – according to early localization attempts. Could the first volcanoes detected on Venus be active ones as well?

A twist of sulfur

These early hints of ongoing volcanism on Venus were strengthened by one additional and somewhat controversial observation in the early eighties. During its orbital survey from 1978 to 1986, Pioneer Venus observed a marked drop in the concentration of sulfur dioxide above the cloud deck of the planet – some ten-fold in seven years – and a concurrent dissipation of the high-altitude sulfurous mist.

For volcano enthusiasts, this had to be the mark of cyclical volcanic activity: in their scenario, major eruptions occasionally blasted large quantities of sulfur dioxide into the upper atmosphere – pulses which were then followed by gradual resorption, as the sulfur condensed into fine droplets of sulfuric acid, and precipitated below the cloud tops. Pioneer Venus would have observed such a resorption cycle from 1978 to 1983, as if the probe had reached Venus in the wake of a major eruption.

If one looks up the record of astronomical observations, a similar variation of SO_2 above the Venus cloud deck can also be traced through the spectroscopic surveys of the late 1950s.

Volcanic eruptions caught in the act? Proponents of the theory rightly pointed out that some plinian eruptions on Earth have been detected by just this method – through the monitoring of SO_2 variations in the atmosphere. But in the case of Venus, the argumentation was far from conclusive. Geochemists warned volcano experts that the atmosphere of Venus was complex, and that the observed variations of SO_2 could be explained by a number of other processes, besides eruptions: they could be due instead to chemical reactions in the upper atmosphere; mixing and overturning of gas layers under the influence of strong winds; or to the formation and dissipation of mist and cloud layers at different altitudes, masking and unveiling in turn the underlying SO_2 gas.

The hypothesis of 'obvious', contemporary eruptions on Venus was further undermined by the realization that the celebrated lightning bolts and thunder claps recorded by the Venera and Pioneer Venus probes could also be explained without resorting to volcanism: tracing their origin to the Beta Regio volcanoes proved to be wishful thinking, and a reevaluation of the data showed that no precise region could be assigned to these radioelectric storms. They could be caused by high-altitude phenomena having nothing to do with volcanoes – namely interactions of the strong solar wind with the charged particles of the planet's ionosphere.

In the early 1980s, the volcanic status of Venus thus remained somewhat controversial: there apparently were a few large shields – as viewed by radar in Beta Regio – and the ground photographed and gamma-analyzed by the Soviet probes did indeed look like basalt. But opinions diverged as to the contemporary nature of the activity – or at least the validity of the arguments raised in its support.

So it was with heightened interest that planetary geologists all over the world waited for new data, as the Soviet Union launched six probes off to Venus in rapid succession: Venera 13 and 14 (1982); Venera 15 and 16 (1983); and Vega 1 and 2 (1985).

The basalts of Venus

Venera 13 and 14 performed successful landings less than a thousand kilometers apart, in the plains east of Phoebe Regio.

For the first time, color images were radioed back from the surface. Orange turned out to be the dominant color, due to the high altitude filtering of sunlight by sulfurous clouds and gas. In addition, the otherwise transparent, lower atmosphere was so dense that light was strongly bent, as through a thick lens: this mirage effect restrained the field of view of the cameras to a mere one hundred meters, instead of the one thousand meters originally expected.

In the foreground, the surface of Venus appeared in exquisite detail: dark rocks stretched out in a thin, broken-up layer, looking very much like fine-grained lava. From the rock size, it appeared that the lava flows were 40 or 50 cm thick. The smooth surfaces and curved edges were reminiscent of 'ropey' lavas on Earth, such as the pahoehoe basalts of Hawaiian shields. Colorwise, the orange hue of the cloud-filtered light dominated the landscape, but correcting for the hue indicated that the lavas were dark gray in essence, a typical color for mafic, unoxidized lava.

On the Venera 13 site (fig. 7.5), patches of gray lava were separated by pockets of darker, granular material, as on the previously photographed Venera 10 site. In contrast, the pahoehoe-looking rock on the Venera 14 site formed a cracked, but near continuous sheet with very little soil. One reason could be that the Venera 14 site is younger, and less eroded than other sites. In fact, Magellan imagery today suggests

FIG. 7.5. The ground of Venus, photographed on March 1, 1982 by Venera 13 in the volcanic plains east of Phoebe Regio (7° S, 303° E). The probe's mast is visible to the left. A cracked lava surface is visible in the foreground, with finer sand in the background. X-ray fluorescence analyses showed the lava to be subalkaline (enriched in potassium). In later orbital imagery, the landing area appeared dominated by dark lava plains, bordered by occasional steep-sided volcanic domes. Credit: USSR Academy of Science, courtesy of the Vernadsky Institute and Brown University.

that the probe probably landed on a field of recent-looking flows, extending from a 75 km-diameter volcanic shield.

Both probes seized ground samples for on-board chemical analyses by X-ray fluorescence spectroscopy: the results were a great improvement over the gamma-ray indications of the previous missions since on these occasions most major elements were quantified.

Oxide proportions on the Venera 14 site matched those of *tholeiitic basalts* (see Table 7.1), with 49% silica and a mere 0.2% of potassium – a lesser proportion of alkali than previously estimated on the Venera 9 and 10 sites. These Venus '*tholeiites*' were therefore close in composition to the basalts of our mid-ocean ridges.

On the nearby Venera 13 site, the oxide line-up was noticeably different, with less silica (45%) and a much higher potassium content (4%). On Earth, this would correspond to the fairly rare class of *subalkaline basalts*, also known as *leucitic basalts*, as are found on the slopes of Vesuvius, and in the Eifel volcanoes of the Rhine graben in Germany. When the Magellan orbiter later imaged the Venera 13 site by radar, it appeared that the area did contain an original volcanic landform, named a *corona* (discussed later in this chapter, and in Chapter 8).

Two more analyses broadened our view of Venus chemistry in 1985: Vega 1 and 2 landed on the outskirts of Aphrodite Terra (180° W), nearly at the antipodes of the previous seven probes. The Vega spacecraft did not carry cameras – part of the payload consisted instead of balloons, released on their way down to the surface, and which drifted at an altitude of 50 km for over two days, collecting valuable data on the

Table 7.1 *Chemical analyses of the ground of Venus*

	Venera 13 (7° S, 303° W)	Venera 14 (13° S, 310° W)	Vega 2 (6° S, 181° W)	Earth (tholeiite basalt)	Earth (alkaline basalt)
SiO_2	45.1 ± 3.0	48.7 ± 3.6	45.6 ± 3.2	49.7	45.1
Al_2O_3	15.8 ± 3.0	17.9 ± 2.6	16.0 ± 1.8	16.5	13.4
FeO	9.3 ± 2.2	8.8 ± 1.8	7.7 ± 3.7	8.3	12.7
MgO	11.4 ± 6.2	8.1 ± 3.3	11.5 ± 3.7	9.0	11.5
CaO	7.1 ± 0.6	10.3 ± 1.2	7.5 ± 0.7	13.8	10.4
K_2O	4.0 ± 0.6	0.2 ± 0.1	0.1 ± 0.1	0.1	0.7
TiO_2	1.6 ± 0.5	1.2 ± 0.4	0.2 ± 0.1	0.7	2.0
MnO	0.2 ± 0.1	0.2 ± 0.1	0.1 ± 0.1	—	—
SO_3	1.6 ± 1.0	0.9 ± 0.8	4.7 ± 1.5	—	—
Cl	0.3	0.4	<0.3	—	—
Na_2O	—	—	—	1.9	2.9
Total	96.4	96.7	93.7	100	98.7

Composition of the ground of Venus, as measured by Soviet probes Venera 13, Venera 14 and Vega 2 by fluorescence spectrometry. Percentages are in mass, with error bars (from Surkov *et al.*, 1983). For comparison, chemical averages of tholeiitic basalt (from the mid-ocean ridges) and alkaline basalt are listed. Note the similarity between Venera 14 and Vega 2 analyses with tholeiitic basalt compositions, whereas the data of Venera 13 is more consistent with an alkaline basalt composition.
From Hess P. C. and Head J. W., Derivation of primary magmas and melting of crustal materials on Venus, Earth, Moon, and Planets, 50/51, 57–80, 1990.

composition and properties of the atmosphere. The landing probes concentrated on performing new analyses of the ground – two gamma spectroscopies, and one X-ray spectroscopy (Vega 2, see Table 7.1). Again, the data pointed to tholeiite basalt chemistry, of 'African rift' affinity.

To summarize, of the seven sites analyzed by the Soviet probes, six display basaltic compositions: a tholeiite trend on most sites (Venera 9, 10, 14; Vega 1 and 2); and one subalkaline variant (Venera 13). One exception stands out, the very first analysis performed on Venus. To the credit of Venera 8, in Alpha Regio in 1972, this early run of gamma-ray spectroscopy indicated a high potassium content (4%), and ten times the uranium and thorium contents that would later be measured on other sites.

Either the Venera 8 values are those of an evolved, subalkaline lava – in which case the closest terrestrial equivalent might be trachyte or syenite – or else the data reflects a different type of alkaline mafic rock all together, such as a *lamprophyre*. Interestingly, when Magellan later imaged the Venera 8 site from orbit, it did show a viscous-looking volcanic dome, 25 km in diameter, within the probable landing circle. So it might well be that Venera 8 did land on a dome of unusual chemistry, or on a lava flow erupted from it.

It is interesting to note in retrospect that the five tholeiitic analyses on the ground match up with orbital imagery that shows mostly plains, whereas the two distinct chemistries (Venera 8 and Venera 13) are associated with the more unusual volcanic landforms. But however distinct – lamprophyre, trachyte, or subalkaline basalt – these chemistries still fall short of the extensive enrichment trends in silica and alkali that are possible on Earth, in subduction settings and in volatile-flushed magma chambers.

Calderas and coronae

Amid its salvo of landing premieres, the Soviet Union achieved a remarkable orbital survey of Venus in 1983–84 with Venera 15 and 16. Taking radar imagery one step farther than the Pioneer Venus topographic mapping of the late seventies, the Russian probes sketched true radar images of the landscape in high definition. Swath after swath, Venera 15 and 16 pieced together a detailed mosaic of the northern latitudes, covering nearly 25% of the planet's surface at a horizontal resolution of 1000 to 2000 m, and with an altitude precision of 50 m.

The northern latitudes of Ishtar Terra were scanned in exquisite detail, revealing mountain ranges that framed a central plateau, named Lakshmi Planum. The mountains were fascinating in their own right, since they were the first fold-like ranges outside of Earth, and rivaled in places the majesty of our own Himalayas: Maxwell Montes, in the eastern range of Ishtar Terra, peaks at a record 12 000 m above the reference level.

The central Lakshmi plateau, on the other hand, shows signs of extension, boasting two large calderas in the midst of its 2000 km-wide expanse. Like other formations on Venus, these were named after famous women: the westernmost caldera was christened Colette – in honour of the French writer – and the easternmost caldera was named Sacajawea, after the Shoshone Indian who guided Lewis and Clark through the Rocky mountains (see fig. 7.6).

Both calderas are oval-shaped (80 × 120 km for Colette; 140 × 280 km for Sacajawea), their outside flanks streaked with bright, radial markings over distances up to 300 km – most likely lava flows or rift structures. The calderas are nested at the top of broad, shallow shields, arching 1000 m above the plateau baseline, and consist of concentric rings of crests and troughs, dropping one to two thousand meters down to the caldera floors.

Other structures revealed by the Venera radars are truly unique, and make up a new geomorphic class: named *coronae*, they consist of circular rings of deformed terrane, typically averaging 200 to 300 km in diameter, with a central region rich in volcanic features such as domes and lava flows (see fig. 7.7). Some coronae are smaller than average (down to 100 km), and a few reach upward to 1000 km (Heng-O corona), with one record structure 2500 km in diameter named Artemis Chasma.

As Magellan's altimeter later found out, the topographic profile of a corona is

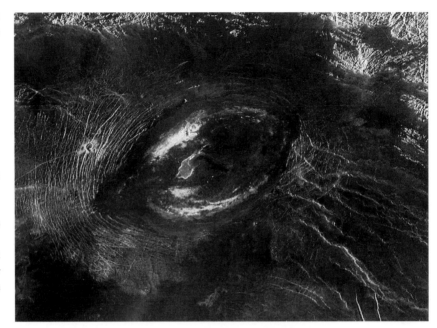

FIG. 7.6. Sacajawea Patera (64° N, 337° E), one of the first volcanic features identified on early radar imagery of Venus. This high-resolution image is a later view collected by Magellan, and shows the elongate caldera (120 km × 215 km) bounded by graben and fault scarps. A flanking rift zone extends to the southeast (lower right), with a shield on its southern margin leading into an oval patch of dark lava flows. An even darker patch of flows is visible in the lower left corner of the image. Bright lava flows are seen inside the rim of the caldera, north and south of the bright central depression. Credit: NASA/JPL (Magellan).

particularly unique. The central arena is often slightly higher than its surroundings, with an assortment of lava fields, flow channels, and shields and domes. It is surrounded by a raised rim, with an annular band of deformation on the outside, and more often than not, a peripheral trench or 'moat'.

Venera 15 and 16 identified some thirty coronae in the high latitudes covered by their survey, clustered in extensional tectonic belts on either side of Ishtar Terra. Coronae were later found to be abundant around Aphrodite as well, and in many other extensional regions of Venus. Today, the corona family encompasses some 300 features.

Venera 15 and 16 also discovered a special type of corona, so distinct as to deserve the separate name *arachnoid* – in view of its web-like lineament pattern. Often smaller and less outwardly volcanic than true coronae, arachnoids are subcircular structures showing both concentric and radial deformation patterns, giving them their spider-web appearance.

Dating the volcanoes

Which ones of these volcanic and para-volcanic features are presently active, if any? With no samples returned to Earth for dating, it is difficult to come up with an answer. The only available dating method consists in counting impact craters on the surface of features to be dated, but scientists run into a problem very specific to Venus. The thick atmosphere of our sister planet is very effective in breaking up and consuming most meteorites before they hit the ground. Only the largest – and rarest – bolides managed to reach the surface, and accordingly only craters larger than 3000 m in diameter are preserved.

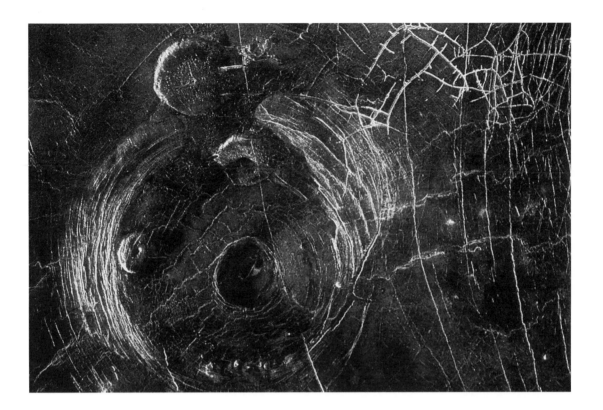

FIG. 7.7. Aidne corona, in the plains south of Aphrodite Terra (59° S, 164° E). The 200 km-wide structure is characterized by a ring of bright ridges and grooves, with associated volcanism in the form of lava flows and domes. Note the large steep-sided dome to the north, and the smaller one inboard of the western crescent of bright ridges. Coronae were first discovered by the Venera 15 and 16 probes and ground-based radars. This high-resolution view is a 1991 Magellan image. Credit: NASA/JPL (Magellan), courtesy of Brown University.

With less than 1000 impact data points for the entire planetary surface, the dating 'grid' on Venus proves to be unreasonably loose, only allowing vague age estimates for very large areas: small features like volcanoes and lava flows 'slip through the net', and cannot be dated individually. Nor will higher resolutions on future missions settle the age issue, since there are no smaller impact craters to be found. Hence, the suggestion that some of the volcanic activity on Venus is recent or even contemporary cannot be answered directly by the crater dating method, and age estimates to this day remain imprecise.

In search of a model

Before the first radar surveys unveiled the surface of Venus, many geologists expected our sister planet to be ruled by plate tectonics similar to our own. Early topographic data seemed to match these expectations.

Aphrodite, the long chain of highlands girdling the equator, was tentatively compared to a mid-ocean ridge, where magma upwelled along the central axis of the massif and was then stretched outwards, cooling and subsiding in the process. This model did find some support in the topographic layout of Aphrodite: in its western half, the chain could be resolved into crest segments striking roughly NE/SW, each segment measuring several hundred kilometers in length, and separated from its

FIG. 7.8. Beta Regio, one of the large volcanic rises on Venus, is cut by deep rifts that supported early theories of wide-scale crustal spreading on our sister planet. This 700 km-wide Magellan image is centered around 33° N, 283° E and shows a north-trending rift extending from Rhea Mons (to the south). The rift is especially visible in the northern half of the image where it is encased by steep walls and bright massifs. Dark patches in the center are smooth lava flows. However spectacular, the rift accomodates only a few tens of kilometers of horizontal extension and is more comparable to the Great African Rift on Earth than to the extensive spreading of our mid-ocean ridges. Credit: NASA/JPL (Magellan), courtesy of Sean Solomon.

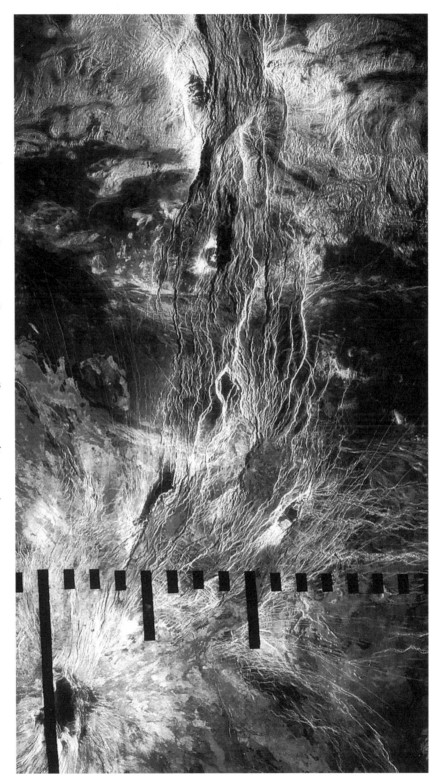

neighbors by NW/SE offsets, that evoked the transform faults of our mid-ocean ridges on Earth.

Reinforcing the mid-ocean ridge model, topographic profiles measured across Aphrodite Terra showed a rough symmetry about its central axis, with declining altitudes on both sides of crest segments, as is the case in ocean-floor spreading on Earth. Taking the comparison one step further, a spreading rate could be calculated for the would-be Venus ridges, by plugging altitude-drop versus distance to the crest into equations of thermal subsidence of a moving plate. Depending on the thermal parameters used, would-be spreading rates at Aphrodite were estimated to fall between 5 mm and 3 cm per year, comparable to plate tectonic rates on Earth.

Early on, it was therefore tempting to model Venus as a plate-tectonic planet, renewing its crust in an earthly fashion. However, for this model to be justified, two confirmations were needed: firstly, did the age of the crust in Aphrodite Terra truly get older, as one moved outward from the central crests and down the side wings, as the model required? And secondly, if spreading did occur, where on Venus was plate subduction, needed to balance out the constant crust renewal at the ridges?

The answer to the first question was tricky. Assigning relative ages to different areas of Venus was nearly impossible, because there were not enough impact craters to get statistically meaningful figures. One could argue that there seemed to be a few more craters in the plains north of Aphrodite than in the presumably younger, central axis, but by no means was the pattern clear-cut and convincing.

As for the second constraint of finding 'subduction trenches' on Venus, there did appear to be a few candidate zones where topography and tectonic deformation were suggestive of converging plate margins, most notably the folded mountain belts around Ishtar Terra. The southwestern belt in particular – Danu Montes – was bordered by a deep trough, the mountain front sloping down at a 20° angle from altitudes over 3000 m to a trench bottom 1000 m below reference level. On the opposite side of the trench, the ground profile sloped up to plain level at a more gentle angle – an asymmetric profile that was characteristic of subduction trenches on Earth.

Although features of the sort were qualitatively supportive of subduction zones and plate tectonics, quantitatively the figures didn't seem to match up. Mountain belts with trench-like margins – essentially the ranges bordering Ishtar Terra – accounted for no more than 3000 to 4000 km of potential subduction zones, versus 10 000 to 20 000 km of postulated spreading ridges in Aphrodite. This amounted to a three to five-fold discrepancy between estimated crust production and crust destruction. So, despite the fact that it was conceptually attractive, the Earth-like model of plate tectonics could not be applied to Venus. The key to its volcanic history lay elsewhere. Finding a suitable answer to this thermal and structural puzzle would become one of the major goals and achievements of the Magellan mission.

8 Magellan explores Venus

The voyage of Magellan

With the Pioneer and Venera probes of the early 1980s Venus' veil had only partially been lifted, so it was with great anticipation that the scientific community awaited the next generation of spacecrafts designed to fully map out the landscape of our sister planet. At the close of the decade, the long-awaited Magellan spacecraft took off for the skies, and after a circuitous route across the Solar System achieved Venus orbit on August 10, 1990, switching on its synthetic aperture radar for the beginning of a two-year global survey.

The images returned by the US probe held up to their promise, with horizontal resolutions under 300 m – a ten-fold improvement over the previous imagery of the Venera 16 and 17 missions. Planetary geologists were deluged with data: large volcanoes blossomed in all their splendor, and small volcanoes popped out of the background in droves. Tectonic features – lineaments, faults, and folds – came into sharp focus, so dense and complex that mission scientists were quick to describe the planet as a volcanic paradise... but a structural nightmare!

By the Spring of 1991, the first cycle of radar mapping was completed, giving us our first global picture of the geology of Venus, soon followed by review articles and models of Venus tectonics and volcanism.

Confirming the topographic data of the earlier probes, Venus turned out to be a world of volcanic plains, with several continent-size massifs rising above base level. The plains account for 80% of the total surface, in an altitude bracket running from −1000 m to +1000 m. They can be classified according to the extent of their tectonic deformation: smooth plains and lobate plains, often resolved by the radar imagery into overlapping, volcanic-looking flows; and reticular plains, buckled up into low-amplitude crests and troughs by subparallel ridges.

The plains were apparently volcanic in origin, in various stages of tectonic deformation. In the northern latitudes, the Venera spacecraft had already identified over 20 000 small shields, a few kilometers across. Planet wide, Magellan brought the count to hundreds of thousands of features, complete with lava flows and lava channels. Many of these small shields and cones stood out in clusters, reminiscent of our own volcano fields on Earth – the shield fields of Iceland, and of the Snake River Plain of Idaho; and the cone clusters of Arizona and northern Mexico, not to mention the innumerable cone fields of our oceanic rifts and abyssal plains.

Size distribution of volcanoes

This abundance of small volcanoes is a familiar pattern, with the number of volcanoes dropping off sharply with increasing size. We find a similar size distribution on Earth and on Mars. Small cones and domes are the dominant volcano population on our planet – both underwater and on land: they are offshoots of shallow magma reservoirs, with a low hydrostatic head. Medium-sized volcanoes on Earth reflect special tectonic environments, such as intersecting, wide faults tapping deeper magma sources. As for the rare, giant volcanoes on our planet, they are associated with hot spot upwellings from depth, breaking through an uplifted, fault-widened crust.

Likewise on Mars, we observe a size distribution boasting few giant volcanoes (the four Tharsis shields, and half a dozen large paterae); some twenty odd intermediate-sized features; and perhaps several thousand small cones and domes, barely identifiable at the resolution level of the Viking imagery.

Prior to the Magellan mission, the volcano size distribution on Venus was estimated at 50 giant volcanoes over 100 km in diameter (versus only one volcanic complex of this size on Earth – the Big Island of Hawaii); 800 large volcanoes in the 100 to 20 km range, comparable in size to Mount Etna, or the African Rift shields (a minimum number, limited to the northern latitudes covered by the Venera probes); and an estimated 22 000 small volcanoes in the same area, from 20 km down to the 5 km cut-off resolution of the Venera imagery.

Not only did Magellan extend the coverage to the entire surface of Venus, but by increasing the resolution to below 300 m, the probe extended the survey down to features a mere 1000 m in diameter (ten pixel points), tallying all features comparable in size to cinder cones and small abyssal seamounts on Earth.

Small volcanoes on Venus

Small volcanoes on Venus appear to display typical shield profiles, often crowned by a summit pit crater (occasionally two); with low-incline slopes; and aprons of lava or ash, spanning a range of radar reflectivities from bright, rough terrain to dark, smooth patches.

Departing from a typical shield profile, less than 10% of the small Venus volcanoes have a truncated cone appearance, with flattened tops, that resemble the table mountains of Iceland, and underwater seamounts. Other exceptional shapes include true cones and domes.

The clustering of many of these small volcanoes into finite fields, 100 to 200 km in width, was one of the many interesting discoveries of the Magellan mission. A volcano field on Venus typically comprises on the order of a hundred features, ranging from 1 to 10 km in diameter (see fig. 8.1).

A total of 646 volcano fields have been identified so far: they occur for the most part in the plains, but are also found on the flanks of large volcanoes (resembling the 'parasitic cones' of terrestrial shields); as well as inside the arcuate ridges of *corona* structures. These 646 fields represent a total of over 50 000 volcanoes larger than 1 km in diameter, and if we extrapolate to the smaller sizes that escape radar detection at current resolutions, the numbers certainly reach into the millions – as they do on our very active ocean floors on Earth.

This clustering of volcanoes into fields 100 to 200 km wide tells us something about magma circulation in the Venus mantle and crust. One hundred kilometers

FIG. 8.1. Small volcanoes on Venus. The cone-shaped mounds, 1 to 10 km wide and 200 to 300 m high, are clustered in a field of 200 features in the plains south of Meshkenet Tessera (65° N, 114° E). Volcano fields are probably caused by shallow hot spots spreading under the surface of Venus. Many individual features have summit pit craters. Orthogonal lineaments criss-cross the area. Credit: NASA/JPL (Magellan).

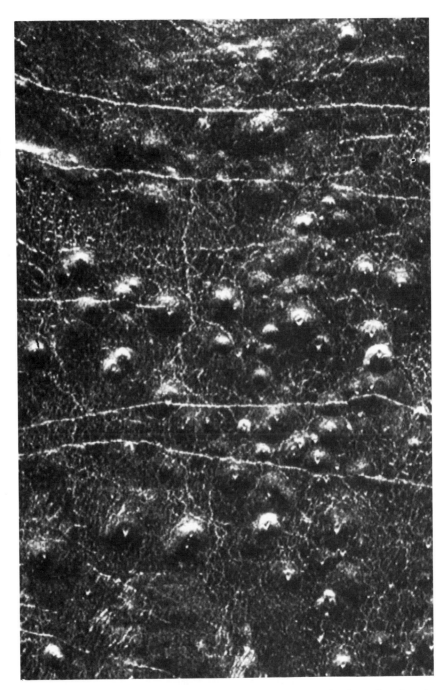

happens to be the estimated width of hot spot plumes as they rise through a planet's plastic mantle. On Venus, the head of a rising column can be pictured as halting several kilometers beneath the surface – upon reaching an equilibrium level where intruding and intruded rock densities balance out – the result being a shallow magma chamber spreading 100 to 200 km in width. Limited tectonic extension, in the ceiling of the bulging chamber, would then establish a network of faults piping small pockets of magma to the surface to build the observed small shields and domes within the perimeter of the underlying chambers.

Large volcanoes

For magma to reach the surface – on any planet – it is necessary for tectonic extension to stretch the crust, and open up access faults. The plains of Venus do show extensional lineaments of the graben type, and large volcanoes are predominantly found in these favorable tectonic settings.

Other terrains on Venus show ridge-like features that are indicative instead of compressional forces. These *tessera* provinces (also known as CRTs: Complex Ridged Terrain) cover 15% of the planetary surface and show very few volcanoes – a direct consequence of the compressional tectonic environment which inhibits the rise and outflow of magma.

The largest Venus volcanoes occupy extensional provinces characterized by abundant graben-like lineaments: Aphrodite Terra, Beta Regio, and Eistla Regio are so torn up by faults that they were compared at first to oceanic spreading ridges (see Chapter 7). Because volcanism on Venus is far from matching the spreading activity of our ocean basins, a closer equivalent might be continental rift zones on Earth, like the East African Rift.

As in terrestrial rift zones, extensional mechanisms at work on Venus are probably linked to the uplift and thinning of the lithosphere over thermal anomalies – hot spots or shallower upwellings in the mantle. Crustal uplift is accompanied by extensional faulting – radial faults near the center of the bulge and concentric faults at the periphery – and those are the faults that buoyant magma can follow to reach the surface.

Large volcanoes preferentially develop at the intersection of several fault planes, resulting in chimney-like feeding pipes. Theia Mons, the tallest feature of Beta Regio, is located at the intersection of at least three rift zones, criss-crossing a regional uplift. East of Beta Regio, volcanoes Sif Mons and Gula Mons are likewise perched on the intersecting fracture zones of the Eistla Regio uplift (see fig. 8.2). As for Ozza Mons in Atla Regio, it lies at the intersection of five chasmata.

Sif and Gula Mons are good examples of large volcanoes on Venus: both are giant shields 300 km in diameter, that owe their magma plumbing system to a dominant set of NW/SE trending faults. Fractures tear through the calderas atop both shields, and extend down the slopes into the surrounding plains: they bring to mind the major rift zones of our own terrestrial shields. Finer lineaments traceable on radar imagery of Sif Mons and Gula Mons are probably subparallel faults, intruded with solidified magma – volcanic dykes.

Calderas atop the larger Venus volcanoes, carved out by summit eruptions and magma chamber collapse, are comparable in size to those atop martian shields: the Sif Mons caldera is 50 km wide, coated with lava fields that appear dark and smooth on radar imagery. Smaller depressions, 3 to 10 km wide, are nested within the main

FIG. 8.2. The merging of radar imagery with altimeter data yields 3-D views of the surface of Venus. This rendition (with a strong vertical exaggeration) shows the volcanoes Sif Mons (left) and Gula Mons (right) in Eistla Regio, with a fracture belt in the foreground. Each volcano is approximately 300 km wide: Sif towers 2000 m and Gula 3000 m above the plains. Credit: NASA/JPL (Magellan), courtesy of Brown University.

floor, an arrangement that is also reminiscent of complex martian calderas (Olympus Mons, Ascraeus Mons), and complex terrestrial calderas (Kilauea for example, with its nested Halemaumau and Kilauea Iki pits).

A family portrait

As of 1994, 167 large volcanoes – greater than 100 km in diameter – have been identified on the Magellan imagery, a three-fold increase over earlier Venera and radiotelescope surveys. Each volcano has a distinct morphology, but a classification can be drawn on the basis of major tectonic and caldera characteristics (see fig. 8.3).

The first two classes regroup all simple volcanoes with radial flows on their flanks and no caldera (Type I), or a simple caldera (Type II). A third category (Type III) comprises volcanoes with both a caldera and a prominent rift zone, as is the case with Sif Mons. Other volcanoes have complex summits: elongated crests (Type IV), exemplified by Gula Mons; or even multiple peaks (Type V) – the showcase example being Sapas Mons with its double summit (see figs. 8.7 and 8.8).

The last four classes regroup those large features that are heavily marked by regional tectonics: those with radial fractures extending away from the shield (Type VI); and volcanoes straddling major, regional rift zones (Type VII), exemplified by Beta Regio's Theia Mons. The last two types comprise those features with central

FIG. 8.3. Classification of
large volcanoes on Venus,
featuring nine different
types from simple shields
(Type I) to shields with
complex summits (Types
II to V) and with major
tectonic deformation
(Types VI to IX). Credit:
From J. W. Head, L. S.
Crumpler, and J. C.
Aubele, 1992, Large
Shield volcanoes on
Venus: distribution and
classification *Lunar Planet.
Sci.*, XXIII, 513–14.

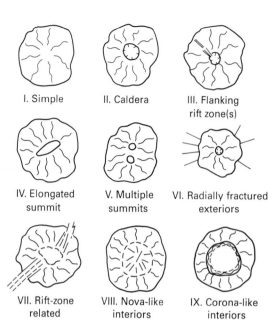

I. Simple　　II. Caldera　　III. Flanking
rift zone(s)

IV. Elongated
summit　　V. Multiple
summits　　VI. Radially fractured
exteriors

VII. Rift-zone
related　　VIII. Nova-like
interiors　　IX. Corona-like
interiors

fault arrays: radial arrays (Type VIII); or concentric arrays (Type IX).

The latter two types are transitional classes, grading to new classes of volcanoes that are specific to Venus and found on no other planet: the *novae*, *arachnoids*, and *coronae*.

Novae are features 50 to 300 km across, carved by a dense array of radial grooves or ridges, which results in a star-like pattern, thus their name. Also called 'stellate fracture centers', the nova family comprises 63 features to date: some are lined up along regional trends; others are perched atop major crustal uplifts and are surrounded by lava fields. Their stellate patterns are most likely caused by magma dykes infiltrating inflated, radially-cracked shields.

Radial/concentric features known as *arachnoids* have even stranger lineament patterns, that resemble spider webs: the central region, 50 to 150 km across, is usually depressed, as if collapsed, and decorated with concentric lineaments, merging with radial lineaments on the outer flanks. Numbering 265 to date, arachnoids are only marginally volcanic, in the sense that they rarely display apparent lava flows. Their lineaments are often more suggestive of compressional ridges than of extensional grabens. Perhaps arachnoids are the topographic expression of underground magmatism, with little lava able to break through to the surface in a compressional environment that seals off incipient fractures.

Another class of original features are the *coronae* – those large circular features first spotted by the Venera orbiters. Coronae are characterized by a circular ring of fractures, often topographically elevated, and sometimes surrounded by a depression or moat (see figs. 7.7 and 8.4). Numbering 206 at last count, coronae range from less than 100 km to over 1000 km in size, with the exceptional Artemis spanning 2500 km – nearly the size of the North Atlantic basin! Coronae are clearly

FIG. 8.4. Idem-Kuva corona on the northern flank of Eistla Regio (25° N, 357° E). Coronae are characterized by a flat to slightly-domed central region, circled with peripheral ridges which focus most of the volcanic activity. Idem-Kuva is 230 km in diameter, flanked by two bright-colored lava fields flowing to the north. The southern part of the corona is torn by a fracture zone propagating toward nearby Gula Mons (just outside of frame). The second circular feature to the left is streaked by a dark patch of deposits, from an impact site on its eastern rim. For a 3-D view of Idem-Kuva, see fig. 8.13. Credit: NASA/JPL (Magellan).

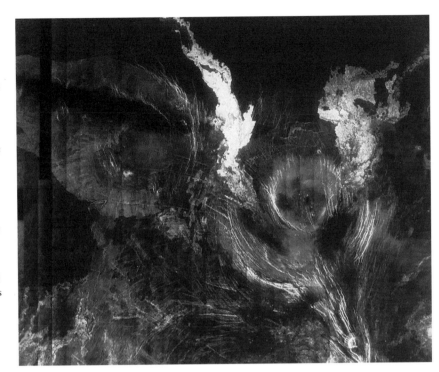

volcano-tectonic in origin, as is apparent from the variety of associated volcanic features: shields, domes, flow fields, and even novae.

Coronae might be the surface expression of upwelling hot spots spreading under the Venus crust, as we shall see at the end of this chapter; although other models call for an opposite regime, whereby some coronae might represent magmatism over downwelling currents in the Venus mantle.

Pancakes and ticks

This family portrait of Venus volcanoes would not be complete without a review of medium-sized volcanoes – those features intermediate in size between the giant shields, novae, and coronae on the one hand, and the small shields and cones on the other. Here as well, Venus graces us with a variety of original volcano types found nowhere else in the Solar System.

Steep-sided domes, also known as *pancake domes*, are one such type. As their nickname indicates, these features are circular lenses of igneous 'dough', with a puffed-up, fractured crust (see fig. 8.5). One is tempted to draw analogies between Venus pancake domes and the siliceous domes on Earth, built of dacite, trachyte, or rhyolite. If the analogy proves valid, this would be the first time siliceous volcanism is discovered outside of Earth. But the steep-sided domes of Venus could just as well owe their viscosity to low-temperature emplacement of a crystal-rich 'mush', with no need for it to be especially siliceous.

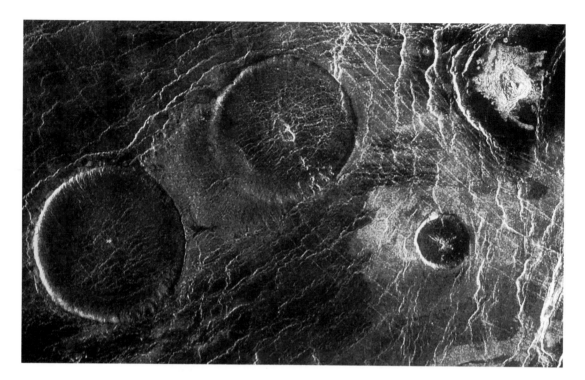

FIG. 8.5. Two large (50 km-wide) steep-sided domes on Venus. Such 'pancake' domes were probably formed by viscous magma stretching a brittle, cooled envelope – perhaps reflecting a siliceous composition. A pearly deposit blankets the saddle between the two domes, spilling over one flank (to the right), as might be expected of a pyroclastic deposit. Note a smaller dome to the right, and a bright impact crater in the upper right corner. Credit: NASA/JPL (Magellan), courtesy of Brown University.

Pancake domes range in size from 20 to 50 km in diameter, and 100 m to 1000 m in height. They number 152 at latest count, and are often found in groups – one spectacular example being the seven domes lined up east of Alpha Regio. In this well-studied array, all seven domes are equal in size (25 km in diameter and 300 m in height), with individual volumes of over one hundred cubic kilometers. This is significantly more than the volume of terrestrial domes, which are rarely larger than a cubic kilometer, but despite their difference in size, Venus pancake domes display the same profile – flattened top, radial summit fractures, and basal concentric fractures – as their terrestrial counterparts. They might share a similar origin: on Earth, 'endogenic' domes are formed by the upwelling of pockets of viscous magma at relatively low temperatures (500 to 600 °C), stretching their cool and somewhat elastic crust into fractured outer shells.

Other original classes of intermediate volcanoes include the modified domes or *ticks*, which display scalloped margins and central depressions (see fig. 8.6); and *anemones*, with radial patterns similar to those of their larger cousins, the *novae*.

Lava flows on Venus

Examples of 'viscous' volcanism on Venus are exceptional: the majority of volcanic formations on our sister planet spill out fluid, basalt-like magma – starting with the prominent lava plains themselves.

Lava plains on Venus spread over the flat landscapes between massifs; in and around *coronae*; outside *tessera* belts; and perched amidst the highlands of Ishtar and

FIG. 8.6. Two intermediate-sized domes, known as *ticks*, east of Beta Regio (18° N, 303° E). These original volcanoes are characterized by scalloped margins and wide summit depressions. The southernmost feature is 45 km across with a 20 km-wide caldera. It is bounded by curved fractures indicating subsidence of the dome following eruptive activity. The surrounding plains are mottled by a variety of lava flows, with contrasting radar reflectivities. Credit: NASA/JPL (Magellan).

Aphrodite Terra. If these lava piles average 1000 m in aggregate thickness (as do *traps* on Earth and *maria* on the Moon), then the Venus plains represent close to a billion cubic kilometers of lava, whereas individual volcanoes like Sif or Theia Mons only measure several hundred cubic kilometers each. Globally then, the plains of Venus might account for over 95% of the planet's volcanic output.

Many lava plains on Venus bring to mind the landscapes of *plains volcanism* on Earth: like the Snake River Plain of Idaho, they are peppered with small shields and short channels, and probably result from shallow partial melting directly below the crust. But in other instances, the lava appears to originate from distant sources outside the lowlands, with long channels carrying the lava down into the plains.

The larger flows are usually restricted to the vicinity of giant shield volcanoes, and many can be traced to the summit calderas and the fracture zones across their flanks. These large flows appear prominently in radar imagery because of textural contrasts with the surrounding terrain, or because distinctive chemical and dielectric properties affect their radar backscatter.

One interesting example of radar signal interpretation is the lava cover at the summit of tall volcanoes, namely at Gula Mons (4000 m) where radar reflectivities are extremely high. This bright signal is believed to be caused by a chemical 'varnish' on the lavas, due to the alteration of minerals by atmospheric gases at those altitudes.

Three alteration minerals are known to give high radar reflectivities of the kind observed atop Gula Mons: *hematite*, *magnetite*, and *pyrite*. The first two are iron oxides, which are not expected to form in the reducing environment of the Venus atmosphere. Pyrite, on the other hand, is a reduced iron sulfide (FeS_2) – a promising candidate since sulfur is abundant both on the ground and in the atmosphere of

FIG. 8.7. Sapas Mons (9° N, 188° E) is a shield volcano rising 2400 m above the surrounding plains, with a double summit (dark central features). The 400 km-wide construct is perched atop the regional uplift of Atla Regio, and displays a complex array of contrasting lava flows (see map, fig. 8.8). The apron of peripheral flows to the southeast (unit 3 on the map) is darker than flows farther upslope – probably due to smoother lavas. To the northeast, one bright flow is seen to partially overrun the ejecta blanket of a circular impact crater. Credit: NASA/JPL (Magellan).

Venus. Pyrite is the familiar 'fool's gold' of mineral collectors, and one can imagine the golden spectacle of glistening lava fields capping the summit of Gula Mons – a patina acquired after years of exposure to the hot, sulfurous gases.

On the lower flanks of a volcano, contrasts in radar reflectivity are probably more a result of surface roughness and block size than chemical make-up, and might indicate a variety of lava flow textures.

Sapas Mons is an interesting case study (see figs. 8.7 and 8.8). It is one of the larger volcanoes on the eastern wing of the Aphrodite highland belt: the shield is over 400 km wide, and rises to a double summit, 2400 m above the plains. The lava flows near the double summit are short and dark: they overlie much brighter, probably rougher flows stretching 100 km down the flanks. The lower portion of the shield is

FIG. 8.8. Map of Sapas Mons volcano (see Magellan image, fig. 8.7). The lava cover has been subdivided into six units, based on contrasting radar reflectivities and super-position relationships – from oldest in the northwest (1) to youngest near the top (6). Bright scarps bound the western edge of the summit mesas, and concentric fractures rim the complex to the east. Parasitic cones are concentrated on the southwestern flank. Credit: From S. T. Keddie and J. W. Head, Sapas Mons Venus: sequence of events in a large shield volcano, abstracts from the 23rd Lunar and Planetary Science Conference, pp. 669–70.

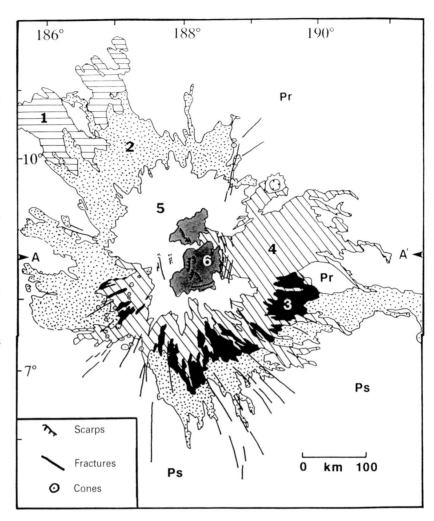

made up of narrower and darker strips of lava (possibly smoother in texture), which average 5 to 10 km in width, and up to 200 km in length.

If we interpret these overlapping series of flows as indicating an evolution of the magma over time – the lower flows being the oldest, and the covering aprons being the youngest – then eruptions on Sapas apparently progressed from fluid flows building up the base of the shield to rougher, perhaps 'clinkier' lavas of slightly higher viscosity. Finally, the stubby flows around the summits might be differentiated lavas, enriched in silica, or basaltic flows of such low discharge rate that they quickly congealed around their vents.

Lava fields and channels

Where large eruptive rates are unleashed, lavas on Venus can spread out in fields as large as the *traps* of Earth, and the *maria* of the Moon. Large flow fields on Venus, in excess of 50 000 km², can be counted by the dozen. Most are associated with the great fracture belts – Aphrodite Terra, Beta Regio, and Atla Regio – others are connected with *coronae*; others yet with individual giant shields like Maat Mons, Sekmet Mons, or Sapho Patera.

One of the most impressive lava flows on Venus, Mylitta Fluctus, originates from a shield straddling a fracture zone in the southern hemisphere. The long flows of Mylitta run down slopes as gentle as 0.1°, and fan out at their extremities in 'deltas' tens of kilometers wide. Long run-out distances on gentle slopes imply very low viscosities – comparable to those of lunar basalts and terrestrial *komatiites* – as they do large eruptive rates. For Mylitta Fluctus, flow-modeling equations indicate rates in excess of 100 000 m³/s.

Funneling these lava flows on Venus are conspicuous channels, which are divided into three categories.

Shallow distribution conduits, flanked by built-up walls of congealed lava (*levees*), run down the axes of many medium-sized lava flows, as they do on Earth. A second type comprises the large and deep sinuous channels, similar to lunar rilles and lava tubes on Earth. Many of the Venus rilles show the same 'tadpole' pattern as their lunar counterparts, with an upstream basin at the vent and a narrowing tail.

Lastly, Venus displays very long *canali* of constant width and depth, meandering over hundreds of kilometers (see fig. 8.9). Canali-type channels reach extreme lengths: one showcase example in Guinevere Planitia exceeds 1000 km, close to ten times the longest lava flow on Earth. Even more extraordinary is channel Hildr, the record-holder on Venus, which extends over 6800 km – the length of the Nile. Hildr has its source near the Atla Regio uplands, close to the equator, and flows northwesterly to the lowland plains of Atalanta. The length of the channel is baffling: even if we assume a flow rate on the order of a million cubic meters per second – comparable to that of the Amazon river – the most fluid lava should have cooled and crept to a halt 2000 to 3000 km from its source. That Hildr reached 6800 km testifies to extraordinary flow conditions, not only in terms of magma output, temperature, and fluidity, but also in terms of heat insulation along the way – implying lava tubes or other forms of thermal protection.

Lava cooling on Venus

Looking at the clear radar images of Venus, we come to forget the presence of the formidable atmosphere, 10 MPa or so of hot gas weighing down on the surface. One would expect this 700 K fluid to act as a warming blanket over the lava erupted at the surface, slow down the cooling rate, and contribute to the long run-out distances of lava flows on Venus. But at the high temperatures at which mafic lavas erupt (1500 K to 1300 K), the 700 K environment is of little significance. On the contrary, the atmosphere on Venus might even speed up the cooling of erupted lava: because of their great density, the gases have a very high potential of heat storage and transport, convecting great amounts of energy away from exposed lava flows. Other conditions being equal, calculations show that the dense atmosphere on Venus should cool

FIG. 8.9. Giant lava
channel on Venus, snaking
along the ridged tessera
north of Freyja Montes.
The 2 km-wide channel
takes its source in the
mountainous cleft to the
right and plunges 3 km
to the dark plains (left),
where it feeds the lava
fronts. Canali of the sort
were probably caused by
extremely fluid magma,
erupting at high temper-
atures and flow rates.
Credit: NASA/JPL
(Magellan), courtesy of
Brown University.

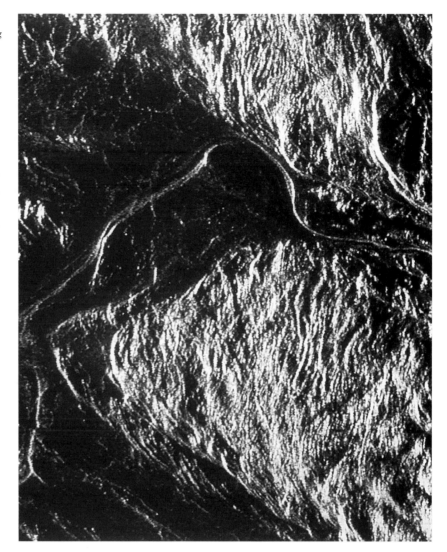

molten lava flows 1.5 times faster than our colder but thinner atmosphere cools lava
flows on Earth.

Surprisingly, high atmospheric temperatures on Venus have perhaps a greater
effect on magma behavior underground than at the surface, since temperature
increases with depth by some 10 K or 20 K per kilometer, depending on which
thermodynamic model one follows: at 30 km depth, temperatures at the base of the
crust might reach 1200 K, and magma rising through fractures will lose little heat by
conduction during ascent, much less than in the colder crust of the Earth and Moon.
Another consequence is that magma will have an easier time melting and assimilating
already hot wall rock along the way, and we might then expect more chemical
contamination and variability of those magma batches that make it up to the surface.

Lava suites on Venus

What compositions might we expect for lavas erupted on Venus? To answer this question we must first define what type of magma can be derived at depth from the partial melting of the Venus mantle rock, and then examine the differentiation trends that the magma might follow during ascent and during stagnation episodes in subsurface magma chambers.

Geochemists believe that the mantle of Venus should not differ much in composition from the Earth's, since both planets accreted in neighboring regions of the Solar System. The main differences seem to be a slightly lower iron content for Venus (our sister planet has an overall density of 5.24 versus 5.52 for the Earth); and mantle temperatures might be higher by 50 to 100 K (although this is highly speculative).

Under these conditions, the mineral assemblage of the Venus mantle should be peridotite-like – similar to Earth's – and high degrees of partial melting in the upper mantle should yield quartz tholeiite and komatiite. Magmas generated at greater depths and higher pressures should again be komatiite (enriched in iron and magnesium); olivine tholeiite; or iron-rich picrite. These predictions are in accord with the fluid appearance of the lava flows imaged by radar; the aspect of lava rocks photographed on the ground; and most of the chemical analyses conducted by the Venera and Vega probes.

Starting with parental magmas of these compositions, how can one obtain differentiated, more siliceous magmas, like those assumed to make up the pancake domes and other viscous types on Venus? Two mechanisms come to mind: the decanting of magma batches in subsurface magma chambers; or else the remelting of the igneous upper crust of Venus, yielding a second generation magma.

The former – the decanting of magma near the surface – will differ from the terrestrial case since there is little, if any, water in the Venus crust to steer the process. The differentiation trend will not be the typical calc-alkaline suite that characterizes 'wet' magma on Earth. Instead, crystal fractionation in Venus magma chambers will drive residual liquids toward tonalitic compositions – lavas rich in silica – if the operation occurs at high-pressure depths around 50 to 70 km; and ferrobasaltic compositions if decanting occurs closer to the surface. Both lava types are fluid.

The second process – remelting of first generation tholeiite or komatiite lava – would necessitate a subduction-like mechanism to drag slices of crust down into the hot mantle. On Earth, the recycling of crustal plates at subduction zones is widespread, with the resulting calc-alkaline suites of andesites and dacites oozing up behind the trenches. On Venus, subduction and remelting of igneous, dry crust (of eclogite composition since the high pressures at depth would compact mineral assemblages) is expected to yield ferrobasalts, andesitic basalts, and andesites if melting is pervasive; and more siliceous lavas of the trondhjemite family if the melting is partial, and limited to the most fusible elements.

These predictions are based on the assumption of a dry, volatile-poor mantle for Venus. If we allow a little water into the system, we would obtain a vapor-rich, siliceous froth of dacitic composition – the type of lava that makes up island-arc volcanoes on Earth. This is an unlikely variation, however, since Venus is believed to contain very little water. Carbonates, on the other hand, are probably widespread on Venus, and would affect volcanism, carbon dioxide driving the melt to trachyte and phonolite compositions – siliceous lavas that are excellent candidates to explain the pancake domes on Venus, and other viscous-looking extrusions.

Eruptions under pressure

In summary, magmas on Venus are expected to fall into two broad categories, depending on their tectonic environment: a first generation, mantle-derived family, with mostly tholeiite and komatiite, and minor amounts of their differentiation products tonalite and ferrobasalt; and a second family in crust-recycling environments, where ferrobasalt, andesite, and trondhjemite (also called leucodiorite) might take the lead, accompanied by their differentiation products trachyte and phonolite.

If we cross-check these theoretical predictions with the spectroscopic analyses of the Venera and Vega landing probes (see Chapter 7), it appears that the 'ground truth' favours first-generation, mantle-derived lavas: all sites are tholeiitic to subalkaline, with the exception of Venera 8 in Beta Regio, which landed near a pancake dome and suggested a tonalite-like composition.

Under this range of compositions, what are the chances of observing explosive eruptions on Venus, such as those postulated to explain the sulfur dioxide fluctuations in the upper atmosphere?

Plinian or pelean eruptions on Venus would be hard to achieve: magma disrupts into fragments only if it is volatile-rich and experiences a severe pressure drop as it reaches the surface. This is hardly the case on Venus, where crushing atmospheric pressures of 8 MPa in the plains, and up to 10 MPa in the basins are comparable to the confining pressures experienced by our underwater volcanoes at depths of 800 m to 1000 m. Even the tallest volcano on Venus, Maat Mons, still experiences 5 MPa pressures at its summit altitude of 8000 m, corresponding to a 500 m water depth on Earth.

Under such high pressures, the volatiles of erupting magmas are not able to decompress much, and the exsolved bubbles remain very small, causing little disturbance. We see this in basalts sampled on our ocean floors on Earth, where vesicules in the rock fabric are less than a millimeter in size. For magmas to disrupt on Venus, very large quantities of dissolved volatiles would be necessary: only then would the sheer amount of gas build up to total volumes sufficient for the magmatic froth to blow apart.

What volatiles could possibly be stored in the Venus mantle in proportions large enough to trigger explosive behavior at the surface? We can take our clues from the Venus atmosphere itself, which owes its composition to billions of years of volcanic outgassing. Carbon dioxide is by far the dominant gas on Venus (96%); followed by nitrogen (3.5%); and only traces of other gases, including less than 0.01% of water vapor.

Based on these observations, water can be ruled out as a major volatile in Venus magmas, in contrast to the situation on Earth where water reigns supreme, affecting melting mechanisms, differentiation trends, and explosive eruptions. Carbon dioxide is more likely to be the dominant volatile in the Venus magma, with sulfur dioxide a distant second. As for the amount needed to drive explosive eruptions at the surface, calculations call for mass concentrations of at least 5% of carbon dioxide (by mass) for eruptions in the lowlands, or 2% in the highlands where the atmospheric pressure is lower. These are high figures to expect on any planet: magmas on Earth harbor only a fraction of one percent of carbon dioxide – about one tenth the amount. Moreover, for a magma to charge up with 2% of carbon dioxide, high lithospheric pressures are needed that correspond to source regions at least 60 km deep. Any shallower, the magma is not physically able to load up with the minimum requisite of 2% carbon dioxide.

It is therefore improbable that eruptions on Venus will be explosive – except for deep-source magmas (below 60 km) stocked with an unusually high (over 2%) volatile content. The majority of eruptions are expected instead to be peacefully effusive.

Eruption styles on Venus

In those rare cases where volatiles do reach disruptive concentrations on Venus, eruptions will still remain low-key, based on theoretical considerations.

If we indeed look at the mechanism of an explosive eruption, there are two phases involved: the heaviest lava fragments are ejected ballistically, like shells out of a cannon, whereas the finer ash stays coupled to the gas in a convective cloud that climbs high above the crater if the eruptive rate is great enough (plinian eruptions); or else collapses and flows down the volcano slopes if the eruptive rate is low (pelean eruptions).

It so happens that the density of the Venus atmosphere will inhibit both ballistic and convective processes, as it already does gas exsolution and magma fragmentation in the first place.

Ballistic ejecta are indeed braked by the friction of the dense atmosphere, rise little, and fall back close to the vent. As for the convective columns of gas and fine ash, they are likewise kept from rising by the thick, confining atmosphere, and limited to one half or even one third the height they would reach on Earth – all other conditions being equal.

Again, it should be stressed that a convective cloud has a chance of developing only if the amount of volatiles is great enough to trigger explosive behavior in the first place; and if the eruptive rate is also large enough for a column to start climbing – i.e. several hundred tons of magma per second. For both conditions to be reached together on Venus is very unlikely, and this casts a serious doubt on the validity of the volcanic hypothesis used to explain the sulfur dioxide variations observed by Pioneer Venus 50 km above the ground.

Based on this theoretical modeling, it is difficult to imagine frequent plinian eruptions on Venus, although plinian air-fall deposits are suspected in several areas, namely on the slopes of Bell Regio and in the plains near Artemis Chasma – where

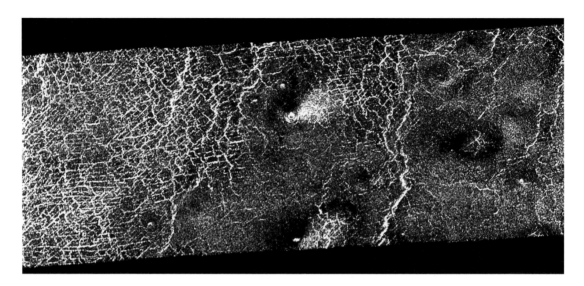

FIG. 8.10. Two small shields with summit pit craters (1 km-wide) in the plains of Guinevere (north is to the left). The bright, web-like reticulate plains in the left half of the image are interrupted by dark plains in the vicinity of the shields, thought to be pyroclastic deposits: from radar signal characteristics, the mantling deposit is estimated to be several meters thick. The bright wedges extending from the shields toward the right are interpreted to be outcrops of reticulate plains, stripped of the dark deposits by wind eddies in the wake of the volcanoes. Other shields and dark mantling deposits are seen to the far right. Credit: NASA/JPL (Magellan).

radar-bright areas display some of the characteristics of pyroclastic flow, including topographic control. Small shields with pit craters in southern Guinevere Planitia are likewise surrounded by mantling deposits (dark in this case), assumed to be plinian deposits several meters thick and reworked by the wind (see fig. 8.10).

On the other hand, hawaiian eruptions – characterized by droplets of magma propelled tens to hundreds of meters above a vent – are more likely to occur on Venus than plinian eruptions. Lesser amounts of volatiles are needed, and eruption rates need not be very high.

In fact, the size of fragments propelled in lava fountains on Venus would be greater than in similar eruptions on Earth, and this is where the high gas density on Venus plays a positive role: thicker gases have the capacity to lift heavier loads off the ground. Another consequence of the unique Venus environment is that tossed fragments will cool little during their short 'hop' through the warm atmosphere, and will land half-molten around the vent: this rain of high temperature 'rock batter' will collect in lava pools and run downslope, mimicking effusive lava flows. But again we must keep in mind that hawaiian eruptions of the sort will still be rarer on Venus than on Earth, because of the relatively higher concentration of volatiles needed to trigger the mechanism in the first place.

Incidently, strombolian eruptions escape most of these planetary considerations. This type of eruption occurs when magma squeezes up tight chimneys at low ascent rates, so that exsolved gas bubbles rise faster than the magma itself, gathering in clusters as they sweep through. Instead of a multitude of small bubbles disseminated throughout the magma, without enough force (i.e. volume fraction) to disrupt it, one gets coalesced bubbles that represent large volumes when they finally reach the top of the chimney – blasting blobs of magma above the vent. In the strombolian mode there is no need for high concentrations of volatiles at the onset: the process concentrates whatever is available to periodically reach disruption thresholds. This is a type of explosive behavior well suited to Venus volcanoes.

The rate of volcanism on Venus

The rate of volcanic activity on Venus is a fundamental question. We are struck by the sheer number of volcanoes on Venus: at the latest count, there are 1728 individual volcanic centers larger than 20 km in size (several times the number on Earth), and small features on Venus also seem to outnumber our subaerial cinder cones and submarine seamounts. Is Venus then more active than Earth?

It does not appear so. If we count impact craters on Venus, the global age of the surface appears to be around 500 million years (a very crude estimate, the error bar spreading from 200 to 800 million years, but we will use the 500 million figure for discussion sake). This is a venerable age, in contrast to the relative youth of ocean basins on Earth, which average less than 100 million years. Thus, volcanoes are more abundant on Venus not because activity there is greater, but because volcanoes have gathered there over longer periods of time without being disturbed.

There are two reasons for this efficient preservation of volcanic landforms on Venus, making them so numerous.

The first reason is that erosion on Venus is orders of magnitude weaker than on Earth. On our planet, subaerial features are rapidly broken down by the water cycle (rain, freezing, and thawing) and by wind-driven saltation and sand blasting. These processes are ineffective on Venus. For one, there is no water cycle at the surface: the little water and sulfuric acid that does condense in the upper atmosphere never makes it to the ground because of the stifling temperatures at lower elevations, which revaporize the droplets. As for aeolian erosion, it is also very weak on Venus: the heavy atmosphere is sluggish, and acts more like a protective lid (as does the deep ocean over our abyssal volcanoes) than as an erosive agent.

The second reason for the safekeeping of landforms on Venus is that there are no geological processes presently at work that might recycle the crust, in contrast to the plate tectonics situation on Earth, where subduction swallows up entire ocean floors – volcanoes and all – on timescales of 100 million years.

Thus, the greater number of volcanoes on Venus gives us the illusion of greater volcanic activity, when in fact we are looking at better preservation conditions. Indeed, once we scale down the number of volcanoes on our sister planet to take into account the greater age of the surface, we realize that there are fewer volcanoes built and less lava erupted on Venus per year than on Earth, by perhaps one order of magnitude: the currently accepted figure is 1 to 2 km^3 of new volcanic crust emplaced per year on Venus, as opposed to 20 km^3 per year on Earth.

This 1 to 2 km^3 per year figure for Venus is again a very global average, obtained by dividing the total volume of lavas on the planet by an average 500 million years. The lack of small impact craters makes it difficult to reach finer time resolutions, and figure out if the volcanic rate was a steady 1 to 2 km^3 over the entire timespan, or fluctuated in any significant way over the years.

Steady state and catastrophism

Working with the little data that they have on global surface age and lava volume, geologists are able nonetheless to make a few general observations that have profound implications for volcanism on Venus.

For one, the 500 million year average age of the Venus surface is telling us something important. There seem to be no impact craters older than this cut-off age. Could it be that the 1 to 2 km^3 of lava flows per year have somehow managed to erase all older craters on Venus to yield this apparent age configuration? But if such slow, progressive burial is indeed to blame, we would expect to discover impact craters at many different stages of concealment: younger craters would be intact; middle-aged ones would be lightly breached by a lava flow or two; older ones would be half covered by the accumulation of lava flows over the years; and the oldest craters recognizable would only show a peak or a rim emerging from the lava fields.

This does not seem to be the case: of the 900 impact craters mapped on Venus, close to 90% are untouched by outside flows – i.e. only 10% are embayed. Unless we are overlooking other partially buried craters in the radar imagery, it does not appear that volcanism on Venus – as we know it today – is actively erasing craters. It would rather appear that volcanism on a totally grander scale renewed the entire Venus surface 500 million years ago or more – impact craters and all – and that the crustal 'clean slate' that ensued then resumed accumulating impact craters and volcanic features, building the new landscapes that we witness today.

This model of a two-speed volcanism on Venus – 'high gear' episodes of major crustal renewal, followed by 'low gear' periods of limited activity such as that observed today – finds some justification in the geophysical picture of Venus that has emerged over the last few years.

Some observers believe that the strain and relaxation characteristics of the surface imply that the rigid lithosphere of Venus is thick – perhaps 200 km or more – so that the hot mantle below cannot possibly evacuate the bulk of its heat to the surface by conduction alone through such a thick shell. For a steady state to be reached (i.e. for heat to be evacuated as fast as it is generated by radiogenic processes at depth), major volcanic outpouring would be necessary to pipe the surplus heat up to the surface – substantially more than the 1 to 2 km^3 of lava per year that appears to be the average today.

The thick-crust model then leads to a critical inconsistency. At the apparent, limited rate of volcanic resurfacing today, Venus should see its inner temperatures shoot up – incapable as it would be of evacuating its heat properly to the surface. Today's 'low gear' volcanism would then be a temporary state: in order not to 'melt down', the planet must traverse periods of contrasting 'high gear' volcanism to periodically speed up the cooling of its overheated interior, and bring temperatures down to a more acceptable level.

This model of cyclical behavior is a new, exciting concept for planetary geologists. In times of low volcanism – such as today's – the thick, insulating lithosphere would cause the mantle temperature to slowly rise. Precariously perched over this overheated, buoyant mantle, the denser lithosphere would soon find itself in an unstable position, yearning to rupture and sink through the mantle.

Once subduction began, it would be expected to proceed swiftly, on account of the low viscosity of the overheated mantle. Slabs would sink at rates of 20 to 50 cm/year, nearly ten times faster than subduction rates on Earth today. At such a pace, the ruptured lithosphere of Venus might completely vanish – be turned inside out, so to speak – on timescales of 100 million years. During these global subduction events, back-arc spreading would resurface the planet behind the subduction trenches, emplacing extensive flood basalts.

Is there any evidence on Venus for incipient subduction today, in a model where 500 million years of 'low gear' volcanism might have brought the planet to the verge of a new global turnover?

As we saw in the previous chapter, the dipping topography bordering western Ishtar Terra might indicate an incipient trench, as might some of the major, circular coronae. Although the latter are thought by many to be the surface expression of upwelling zones, a few do show an uncanny resemblance to subduction arcs on Earth – namely Artemis, Latona, and Eithinoha coronae, which resemble trench features both in planform and cross-section. Artemis in particular has been compared to the Pacific Ocean's South Sandwich subduction zone. Could these outstanding coronae be the first 'trap doors' of a new global phase of subduction?

This catastrophic scenario of a planet with a periodically thickening crust, leading to mantle overheating and global overturn is but a model. Alternatively, one can interpret the geophysical data as indicating that Venus is a thin-crusted planet, evacuating its heat in a balanced, steady-state fashion. No pulses of plate tectonics would take place, but slices of lower crust would occasionally 'peel off' to sink in the mantle, balancing the low rate of crustal formation.

In this steady-state model, mantle convection would drive broad upwellings to form volcanic rises, rifts and coronae at the surface, while downwelling would occur beneath the crumpled plains, delaminating slices of the lower crust, and letting warm mantle rise in their place to drive plains volcanism (see fig. 8.11).

One could then look at the age configuration of the Venus surface – the apparent resetting of the impact crater 'clock' 500 million years ago – as simply marking the end of major tectonic and volcanic resurfacing on Venus, when global activity simply slowed down as a result of the secular cooling of the planet.

Organized volcanics on Venus

Whatever happened in the past, there is no strong evidence today for large-scale crustal spreading or subduction on Venus. The distribution, shape and size of volcanic features are more reminiscent of hot spot volcanism, and it is hardly a coincidence if the estimated output of 1 to 2 km^3 of magma per year is comparable to that of our own hot spots on Earth.

The distribution of the 1728 volcanic centers on Venus larger than 20 km

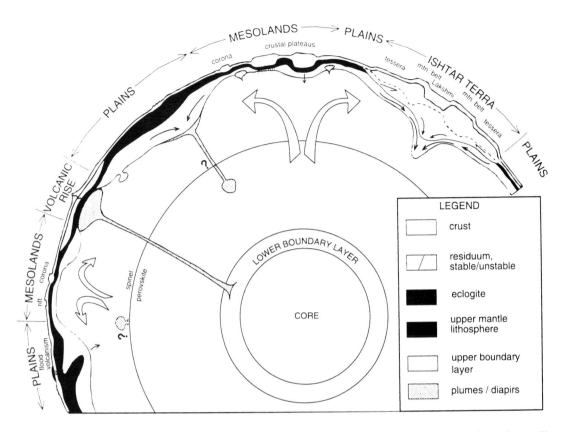

FIG. 8.11. Cross-section (not to scale and highly schematic) of Venus tectonics and volcanism. In this model, mantle upwelling occurs beneath the mesolands, creating coronae and crustal plateaus (top). Downwelling would occur beneath the plains, occasionally triggering delamination and the replacement of subducted crust by warm mantle, giving rise to flood volcanism in the plains (lower left). Volcanic rises would be due to hot spots from the lower mantle. Credit: From R. J. Phillips and V. L. Hansen, 1994, Tectonic and magmatic evolution of Venus, *Annu. Rev. Earth Planet. Sci.*, **22**, 597–654.

confirms this likeness to hot spots. As on Earth, the features are concentrated in two global zones on the planet, where the numbers surge from an average of two or three volcanic centers per million square kilometers to approximately twice or three times that number: one broad area 13 000 km in diameter, framed by the Beta, Atla, and Themis regions – and named for this reason the BAT zone (centered on the equator at 250° E); and a more patchy zone in the opposite hemisphere, with volcanic centers dispersed about a center at 70° E. This is very similar to the Earth's hot spot-intensive hemisphere (Atlantic–Africa–Indian Ocean), and opposed patchy hemisphere, although the numbers of features on Venus are about ten times those on Earth, on account of the good preservation of landforms as we mentioned earlier.

This distribution pattern of hot spots on Venus is therefore geometrically simple, with one dominant 'upwelling' hemisphere of fracture belts, coronae and rifts (the BAT area), surrounded by narrower downwelling zones – plains, tesserae and ridge belts – where large volcanic centers are rare.

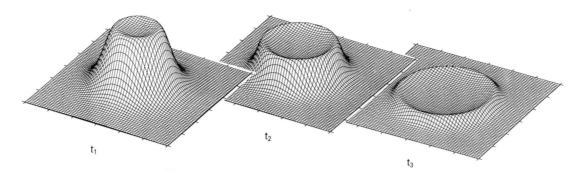

t_1 t_2 t_3

FIG. 8.12. How to form a corona on Venus: cartoon model of evolving topography, as a plume impinges beneath the lithosphere and spreads outward. In the initial stage a broad, dome-like uplift is produced (t_1). As the plume spreads radially beneath the surface and cools, the surface subsides and grows outward into a corona-like feature (t_3). Credit: After D. M. Koch and M. Manga, 1995, A model for corona topography: the ascent and spread of a diapir at a depth of neutral buoyancy, in press.

Hot spot growth on Venus

As our understanding of Venus sharpens, models are drawn that explain volcanic features as successive stages of hot spot growth and maturity – with the plumes impinging on the lithosphere and spreading beneath it.

The deeper, larger plumes would create volcanic rises such as the Beta, Atla, Bell, and Eistla regions. These are characterized by radial rifting and large shield volcanoes, but lava fields are relatively limited as can be seen on the Beta Rise where patches of old cratered terrain are still preserved. This might be due to the fact that deep plumes send a limited amount of magma to the surface, or else that the melt remains trapped at the base of the lithosphere for some reason.

Coronae are smaller and more numerous than volcanic rises and could represent plumes originating at lesser depths, probably at an upper mantle/lower mantle boundary. Simulations (see fig. 8.12) show that a plume reaching the rigid lithosphere will stretch it into a domical uplift, cracked by radial faults, and that with time the hot spot pocket will spread laterally under the lithosphere, cooling and causing the overlying topography to subside: during this maturing phase, the surface will undergo compressional, concentric deformation, as the underlying stress source moves radially outwards.

In matching features on Venus to this 'spreading drop' model, a nova volcano could be the early expression of a hot spot uplifting the lithosphere into a dome, with a radial, star-like pattern of extensional faults.

The mysterious arachnoids, with both radial and concentric lineaments, would illustrate the next stage of plume evolution, when it begins to spread outward, and superimposes concentric strain onto the initially radial fault pattern. Interestingly, observations seem to indicate that the larger the arachnoid, the more concentric features seem to dominate over radial ones, as one would expect from a size-dependent switch from radial to compressional regimes.

The third stage of plume spreading would lead to a corona morphology, with lesser topographic relief as the plume thins out and cools, a near absence of radial features, and a now distinct ring of compressional ridges over the slowly-moving front of the subterranean 'wave'. In support of this interpretation, double-ring coronae do show volcanism to start first in the inner ring, followed by faulting and volcanism in the outer ring.

To summarize, a 100 to 150 km-wide plume would begin by creating a nova,

FIG. 8.13. Idem-Kuva corona in 3-D perspective (compare with the plan view of fig. 8.4), with an associated rift zone (the prominent bright fractures in the foreground). Coronae are believed to be the surface expression of hot spots, spreading radially outward beneath the lithosphere. Magma rises through fractures in the expanding ring of deformation: a bright lava flow can be seen in the background (left of center) flowing north toward the horizon. Credit: NASA/JPL (Magellan).

which would then grow and subside into an arachnoid, before ending its career as an even wider and flatter corona. As for the timescale of this evolution, plugging numbers into the model yields periods of 200 million years for the expansion of a nova into a corona.

Large, complex regions might be the result of a superposition of several hot spots of different ages and stages of growth. The complexity of Beta Regio could be due to a succession of plumes affecting the same spot repeatedly. And the elongated plateau of Aphrodite Terra could represent a line-up of spreading plumes of different ages, with the youngest plume – and highest uplift – in the east (Tethis Regio); an older spreading plume with lower topography to the west (Ovda Regio); and the oldest, lowest feature in the south – the giant, flattened corona of Artemis.

As the analysis of Magellan data continues, these exciting new theories will continue to be refined and challenged, but already it appears that our sister planet has a volcanic character very much its own, both in the way it lives through megacycles of contrasting volcanic activity – if the theory of catastrophic overturn proves correct – and in the way it continually stretches and modifies its hot-spot features, through a complex pattern of radial spreading and subsidence.

The discovery of Venus is far from over, and the question is still not settled as to how recent the last eruptions there have been. Perhaps in the near future, when we return to our sister planet, will we catch in the act one of those grandiose eruptions. Or perhaps we will have to wait a very long time. But one thing is certain: Venus will beckon us to return.

9 Io and the outer moons

FIG. 9.1. Io, the innermost moon of Jupiter, as viewed by Voyager 1. The complex pattern of dark circular features and bright patches is the result of intensive, permanent volcanic activity. The large heart-shaped feature to the left is the ejecta plume of Pele, the largest eruption monitored during the Voyager 1 encounter. Credit: NASA/JPL (Voyager), courtesy of Cornell University.

On March 5, 1979, 19 months after its launch from Cape Canaveral, the Voyager 1 space probe reached the outskirts of Jupiter, its first call in a long round of visits to the outer planets of the Solar System.

Jupiter is the largest of the giant planets, a massive ball of hydrogen 143 000 km in diameter, that could lodge a thousand Earths inside its globe. Fascinating as it may be for planetologists – on account of its chemistry and turbulent atmospherics – Jupiter is of little relevance to geologists: it has no rocky surface, or rather whatever rocky nucleus it might have is buried under thousands of kilometers of gaseous, liquid, and metallic hydrogen.

Fortunately, what Jupiter lacks in geology, it makes up with its harem of rocky moons: four of its satellites are comparable in size to our own Moon (the Galilean satellites, visible from Earth with a pair of powerful binoculars), and another nine, smaller satellites were known prior to the Voyager encounters. Farther out in the Solar System, Saturn and Uranus also have sets of moons; and distant Neptune has one major satellite (Triton).

One of the principal missions of the Voyager program was to photograph and map these rocky worlds, and excitement was at its peak at the Jet Propulsion Laboratory in Pasadena when Voyager 1 swung through the Jovian system, on a trajectory that would take it past the four largest moons: Callisto, Ganymede, Europa, and innermost Io. Voyager's cameras were programmed to collect a wide set of images during the flyby, and began calibration runs several days before the speedy encounter, capturing telephoto imagery of Jupiter's companions.

Eruptions on Io

One moon in particular was begging for attention: Io, the smallest but the most colorful Galilean satellite, shone with a bright yellowish hue, and the Voyager cameras resolved the disk into a wonderful mosaic of yellows, oranges, and reds, laced with patches of green, brown, black, and white. Later color corrections that took into account the various filters used in the imaging (which shifted all colors toward the ultraviolet part of the spectrum) showed Io to be predominantly pastel yellow, laced with oranges, greens and grays.

Besides its outstanding color palette, Io also featured distinct circular patterns: thought at first to be impact craters, they were soon made out to be volcanic calderas. At resolutions close to 1 km, colored flows were identified, emanating from fissures and calderas. But a bigger surprise was yet to come.

As Voyager 1 sped past Jupiter, the spacecraft's trajectory was closely monitored by the navigational team of the Jet Propulsion Laboratory, for the plan was to use Jupiter's strong gravitational field to sling the probe toward distant Saturn – Voyager's next objective in its multi-planet mission. In order to control their gravitational ricochet, navigators at JPL pored over the images transmitted by the spacecraft, using them to compute the probe's trajectory with respect to the background star field, moon positions, and other Jovian reference points.

It was with this job in mind that JPL navigator Linda Morabito and her team analyzed Voyager images of Io recorded on March 8, in order to measure the angular separation between Io's horizon and a pair of reference stars in the dark background. The JPL team had to stretch the picture's contrast in order to pinpoint the faint stars above the moon's disk. This led to a dramatic discovery: brought into light by the contrast adjustment, an umbrella-shaped aura blossomed over Io's horizon. A billion kilometers from Earth, Voyager had recorded the first volcanic eruption ever observed on a planet outside our own.

The first eruption spotted by Voyager 1 was by no means a small one. The umbrella of ejecta rose 300 km above the horizon, and fell back over an area over 1000 km wide. In plan view, the gigantic cupola was seen to be slightly asymmetrical – in the shape of a heart – and appeared to jet out from a dark marking on the surface. The volcanic complex and its plume were named Pele, in tribute to the fiery goddess of the Hawaiian islands.

As more detailed imagery became available, Pele's surface deposits showed radial spokes stretching 150 km from the vent, perhaps due to jets of coarser ejecta streaming down parabolic trajectories. The dartboard plan view was further enhanced by two concentric rings – one a pale yellow, the other a dark brown – rimming the central structure out to a distance of 700 km. On the perimeter, a bright margin completed the picture. It appeared that the multiple rings of ejecta on the ground might record successive stages in the evolution of the eruption, as particles were sprayed progressively farther out.

The pattern of surface deposits and infrared data showed Pele's main vent to be located on the edge of a triangular plateau 200 km wide, torn by an axial

Table 9.1. *Eruptive fountains on Io*

No.	Name	Lat., Long.	Diameter	Altitude	Gas velocity at vent
P1	Pele	19° S, 257° W	1000 km	280 km	930 m/s
P2	Loki	19° N, 305° W	210 km	100 km	580 m/s
P3	Prometheus	3° S, 153° W	250 km	70 km	490 m/s
P4	Volund	22° N, 177° W	75 km	95 km	570 m/s
P5	Amirani	27° N, 119° W	200 km	80 km	520 m/s
P6	Maui	19° N, 122° W	250 km	80 km	520 m/s
P7	Marduk	28° S, 210° W	180 km	120 km	630 m/s
P8	Masubi	45° S, 53° W	150 km	70 km	490 m/s
P9	Loki 2	19° N, 305° W	210 km	100 km	580 m/s
P10	Surt	46° N, 336° W	?	?	?
P11	Aten	48° S, 311° W	?	?	?

Eruptive plumes observed on Io during the Voyager flybys. The plumes are believed to result from volatilization of sulfur dioxide and elemental sulfur jetting through cracks in the uppermost crust. The exit velocities of gases at the vent are estimated from the heights reached by the plumes. Heights and diameters are those initially listed at the time of discovery. Other estimates have been proposed since. Plumes 10 and 11 were not directly observed but were inferred from the change in albedo patterns on the ground that occured between the Voyager 1 and 2 encounters.
Modified from Strom *et al.*, 1979, and Smith *et al.*, 1979.

trough – more precisely on its northern margin, within a black elongated depression roughly 24 km long and 8 km wide. Other vents highlighted by dark pyroclastic deposits were identified on the floor of the graben, and elsewhere along the margin of the plateau. As regards the plume itself, contrast stretching underscored a complex filamentary structure against the dark sky, complete with swirls, which indicated that the ejected particles were interacting with the gas flow.

Pele was not the only region calling for attention. Within a few hours, other plumes were discovered on the processed Voyager imagery. On the wide-angle image showing Pele's umbrella on the horizon, a second eruptive cloud – named Masubi – was spotted on the terminator, its gases illuminated by the grazing sunlight. As the imaging team pored over the Voyager data, another major eruption was identified on eight different frames. It was traced to a fissure zone 180 km long and 20 to 50 km wide, just north of a dark, horseshoe-shaped caldera, and was resolved into two different plumes: a large one at the western end of the fissure and a smaller one at the eastern end. The plumes, fissure, and caldera were named Loki after a Scandinavian god.

The main Loki eruption was smaller and less symmetric than Pele, and consisted of a globular plume 200 km wide and 200 km tall, lacking a distinct central column (later imagery from Voyager 2 showed the plume to surge to heights of 300 km). Interestingly, the Loki plume appeared larger and taller when viewed through the camera's ultraviolet filter. This extended halo was apparently made up of finer particles than the inner part of the plume: their scattering properties indicated sizes in the range 0.002 to 0.02 micrometers, whereas the 'coarser' particles of the central section appeared larger than 2 micrometers.

A third eruption was located a full 100° east of the first two features, close to the equator (whereas Pele and Loki were respectively in the southern and northern tropics). Named Prometheus, this volcano displayed an umbrella-shaped plume 75 km high and 250 km wide, with a dark central fountain (see fig. 9.2). Surrounded by the now familiar bull's-eye pattern of circular markings – a dark inner ring and a white outer ring – the plume displayed radial spokes or 'jets' similar to those of Pele, attributed to funneling effects at the vent or ballistic mechanisms.

Poring over the Voyager images, scientists located several more plumes. The fourth event was a narrow feature 75 km wide and 80 km high named Volund; followed by a pair of wider plumes to the east, less than 300 km apart: Amirani and Maui. A seventh plume, Marduk, was discovered to the southeast of Pele, and an eighth, Masubi, was identified near the terminator. Finally, the eastern plume of Loki, only 20 km high during the Voyager 1 encounter, was labeled plume number 9.

The nature of the propelling gases was determined by Voyager's infrared spectrometer (IRIS): the analysis of the Loki plume showed the dominant gas to be sulfur dioxide. Thermodynamic calculations confirmed that SO_2 was indeed an excellent candidate to be the driving gas of the plumes, although the taller Pele plume required higher SO_2 temperatures, or else elemental sulfur (S). As for the solid particulate matter entrained by the jets, they were calculated to be specks a few micrometers in size (based on their spectral properties), most likely condensed particles of the driving gases.

The extraordinary altitudes reached by eruptive plumes on Io are principally due to the near vacuum reigning at the surface (Io has too weak a gravity to maintain a sizeable atmosphere). Erupting gases expand dramatically at the vent, and reach exit velocities on the order of 500 meters per second. Pele, with its hotter or lighter gases, might even sustain erupting velocities closer to 1000 m/s. This is many times the velocity of erupting gases on Earth, which rarely exceed 100 m/s (except in steam blast explosions, where velocities up to 600 m/s have been measured, but for periods of only a few minutes). Vigorously propelled at the vent, plumes on Io climb to great heights, because of the lack of atmospheric friction on Io and the moon's low gravity.

Tidal energy

Such intense, contemporary volcanism on Io is disconcerting at first. Io is similar in size to our own Moon (3680 km in diameter, versus 3470 km), and one would expect it to be just as extinct today, after having expended its accretional and radiogenic heat in the first two billion years or so of its history. This not being the case, Io departs from the simple, size-related activity curve of terrestrial planets, and calls for an

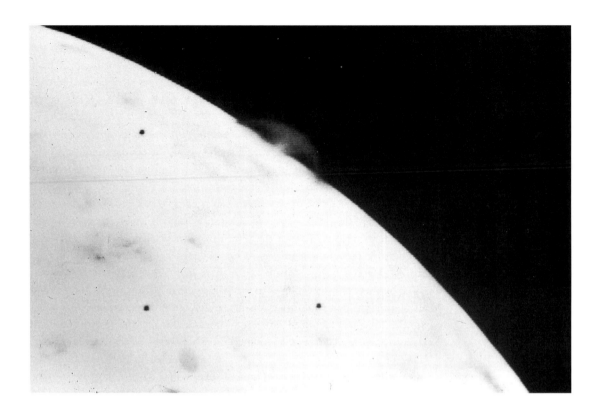

FIG. 9.2. A volcanic plume on Io's limb, viewed by Voyager 1. Named Prometheus, this was the third plume discovered by the probe, and was still active when Voyager 2 flew by Io four months later. The image contrast has been stretched to reveal the structure of the plume: the umbrella of particles rises to an altitude of 70 km and falls back over a diameter of 250 km. Credit: NASA/JPL (Voyager), courtesy of Cornell University.

original heating mechanism. Such a mechanism was actually proposed prior to the Voyager encounter and the discovery of Io's volcanic plumes: calculations had indeed shown that Io might be heated by the large tidal stresses exerted by Jupiter.

Tidal-driven volcanism was a new concept, although tidal forces as a whole were well known. We are especially familiar with tidal phenomena on Earth, because of our relative proximity to the massive Sun, and our close proximity to the Moon: both exert a substantial gravitational pull on our planet. The existence of large, fluid oceans at the Earth's surface turns these forces into an everyday experience: the oceans react perceptibly to tidal pull, swelling and abating as our rotating planet brings them alternately in and out of line with the gravity vectors of the Sun and Moon.

What is less visible is that tidal forces not only affect the liquid cover of the Earth, but also its solid and gaseous envelopes: the atmosphere expands and contracts twice a day as do the oceans, and the rocky crust of the Earth is likewise strained and relaxed. Although solid crust deformation is too slight to be measured by instruments at our current sensitivity thresholds, it can be indirectly inferred: it does seem that some very active volcanoes – such as Hawaii's Kilauea – show bi-weekly fluctuations in their eruptive behavior, as if the stronger tidal pull of the Moon twice a month (spring tides, when Sun and Moon are in line) somehow affected magma reservoirs, perhaps through the rise of the groundwater table around the reservoir, providing steam to the magma (e.g. Santiaguito volcano in Guatemala).

Earthquakes should likewise show a bi-weekly factor in their occurence pattern,

FIG. 9.3. An erupting plume on Io, imaged by Voyager 1. Micron-sized particulate matter – mostly chilled gas – causes the halo. The two-layered structure is due to finer particulate matter in the upper part of the plume. The plume is 250 km wide and over 200 km high. Credit: NASA/JPL, Photothèque Planétaire, d'Orsay, courtesy of Dr Philippe Masson.

although no such factor has yet been isolated in the midst of many stronger tectonic influences. However, it has been observed that the Moon – which feels the tidal pull of the Earth much more strongly than the reverse – does display a bi-weekly cycle in its subdued, seismic activity: Apollo seismometers showed the rare, low-energy quakes deep inside the Moon to be concentrated at times of closest approach to our planet, and farthest excursion away from it, twice a month. It is at those points in its elliptic orbit that the Moon experiences the greatest change in its orbital velocity, whereas its rotation rate remains constant: the combined effect is that the Moon momentarily departs from its synchronously-locked attitude with respect to our planet, and its face appears to 'swing' as viewed from Earth. This relative, swinging motion is known as the lunar libration. Because it occurs within the Earth's strong gravitational field, the oscillation of the Moon's mass leads to stress build-up, released in the form of moonquake energy.

Io and Europa

So it is for the Earth and Moon. Around Jupiter, the tidal relationship between the giant planet and its moons is even stronger. The Galilean satellites are all locked in synchronous, circular orbits, rotating in step with their revolution, and presenting the same side to Jupiter. The near-perfect circularity of their orbits all but eliminates the libration effect, with no variations in gravitational pull. In the case of Io, the orbital eccentricity (departure from a perfect circle) is only 0.00001, in the absence of any outside perturbations, and one would expect the moon to be internally at rest, devoid of any tidal stress.

What Stanton Peale and co-authors calculated, however, was that the Jovian moons were themselves a source of perturbation, tugging and warping each others' orbit as they circled past one another. The interaction between neighbors Io and Europa turned out to be the most disturbing because they were closest to Jupiter, and were subject to the strongest gravity gradient.

The perturbation process works as follows: while innermost Io circles Jupiter in 42 hours and 27 minutes, outer and slower Europa orbits in 85 hours – essentially double the period. As a result, Io overtakes Europa every two laps, the two moons tugging each other away from perfectly circular orbits at each pass: Io's orbit, in particular, is stretched into a slight ellipse with a 'forced' eccentricity of 0.0043. This is a very modest distortion – Io's distance to Jupiter varies by only a few kilometers between periapsis and apoapsis – but because the moon is so close to Jupiter, such a variation has an enormous impact in terms of gravitational pull. Io's rocky body is flexed and strained by an oscillatory libration of high amplitude and high frequency (42-hour cycle). This cyclical stress is dissipated internally in the form of heat, and this is where Io finds the energy to drive its outstanding volcanic activity.

Calculations show that if Io is taken to be a homogeneous sphere of density 3.5, its tidal heating should reach 10^{12} watts – about ten times the radiogenic heat currently produced within our own Moon. But Io's tidal heating would be even stronger if instead of being a homogeneous sphere, it started off as a stratified body composed of a molten core and a solid mantle – as is thought to be the case. One would then expect tidal heating to be concentrated mostly in the asthenosphere at the top of the mantle, and at the mantle/core boundary.

As if Jupiter's tidal heating were not enough, Io is subjected to another, original energy source: 'electric heating', although trivial compared to the tidal effect. It so happens that Jupiter is enveloped in a strong magnetic field (4.2 gauss at cloud-top level), which accompanies the planet in its clipping 10-hour rotation. Orbiting very close to the planet, Io is literally swept by this magnetic field at a relative speed of 57 kilometers per second, and since the rocky moon is reasonably conductive, 600 000 volts of potential is set up across its diameter, funneling currents on the order of 1 million amps: if all this electric current were turned into heat by Io's resistance, up to 2×10^{11} watts could be generated. In practice, however, the diversion of the current by the ionosphere and other factors severely limit this exotic process, so that it probably accounts for only a fraction of one percent of Io's global energy budget.

Calderas on Io

The extent of Io's volcanism is phenomenal, and is not limited to the nine erupting plumes observed by the Voyager spacecraft: a thorough analysis of the moon's disk shows in excess of 200 dark spots over 20 km in diameter (and up to 200 km), each thought to be a volcanic caldera. In fact, the majority of erupting plumes observed on Io are associated with calderas, and probably many more were active at the time of the Voyager encounters but their plumes were too small and too faint to be resolved on the imagery.

FIG. 9.4. Calderas and volcanic plateaus on Io. The caldera in the upper right corner, perfectly circular and pitch black, is about 75 km in diameter. Center frame, a bright deposit – probably sulfur dioxide frost – is nested between a layered plateau (left) and a shield with an elongate caldera and radial flows (right). The frost was probably sprayed out of the plateau by geyser-like activity. Prominent in the lower part of the image is Maasaw caldera, surrounded by an intricate array of dark radial flows. Credit: NASA/JPL (Voyager), courtesy of Cornell University.

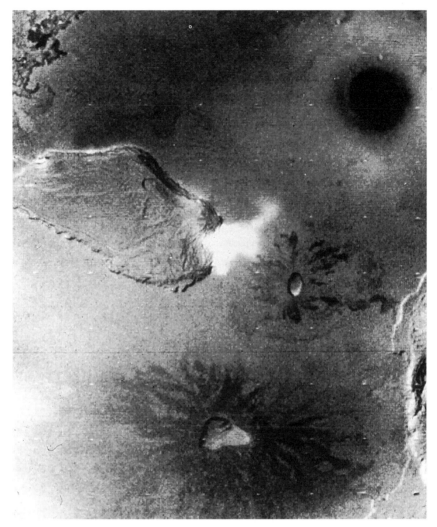

Io's calderas are highly visible because of their contrasting albedo with the surrounding landscape, which makes them appear pitch black in most picture renditions (see fig. 9.4). They are not perched atop giant shields, as is frequently the case on the Earth, Venus, and Mars, but simply dot the plains. Gentle slopes are deduced where lava flow patterns are radial about the caldera, but most often the dark pits are in the midst of plains, or even at the bottom of shallow depressions, as seems to be the case for Loki. The closest terrestrial analogs, from a topographical point of view, would be the calc-alkalic 'ash-flow plains' calderas, such as Japan's Aso caldera and other volcano-tectonic depressions associated with silicic ignimbrite deposits.

Very few volcanic features on Io, then, display significant relief. One exception is Ra Patera, a lava shield several hundred kilometers wide (see fig. 9.5), that slopes up to an altitude of perhaps 2000 m. Two other low, circular shields have been spotted west of Ra Patera – Inachus Tholus and Apis Tholus – and a fourth shield, 2500 m in height, has been identified southeast of Pele. Interestingly, Inachus and Apis appear to be rimmed by basal scarps as is Olympus Mons on Mars, although they are of course much smaller (170 km in diameter and probably 1 to 2 km high).

If most calderas on Io lack built-up pedestals, they often display steep negative relief, dropping abruptly one to two thousand meters to their dark smooth floors. They are circular to elliptical in plan view, and some have scalloped rims that could be due to successive episodes of downdropping – as is common for caldera walls on volcanoes of the Earth and Mars – or some form of volatile sapping. As for the nature of the dark caldera floors, their very low albedo makes it difficult to pinpoint much detail, although stretching the images' contrast does reveal darker than average patches at the foot of some cliffs, interpreted to be the freshest flows, and bright patches thought to be sulfur dioxide frost. In view of the jetting plumes and the overall level of activity on Io, it is indeed to be expected that most caldera floors have recently been resurfaced by lava flows – and that a great number might even harbor active magma lakes, bubbling away under the Jupiter-lit sky.

Volatiles on Io

Infrared data has shown that the driving volatiles of Io's eruptions are mainly sulfur dioxide, along with some elemental sulfur. Io's overall atmosphere, however scant (less than 10^{-2} pascals of pressure), is also dominated by sulfur dioxide.

Sulfur dioxide is a familiar volcanic gas on terrestrial planets. On Earth, it makes up a substantial proportion of erupting volatiles – often 5 to 10%, which puts it in third place behind water vapor and carbon dioxide. Sulfur is abundant on Venus as well: droplets of sulfuric acid make up its clouds, and associated, fluctuating levels of sulfur dioxide are believed by some to reflect episodic plinian eruptions, as discussed in Chapter 7. On Mars, although there are no noticeable sulfurous gases in the atmosphere at present, a significant amount of solid sulfur does show up in the soil (7%), tied up in mineral compounds thought to be essentially sulfates (see Chapter 6).

Therefore, it is not surprising that sulfur dioxide plays a starring role on Io. Nor is it surprising to find virtually no other gases: only heavy molecules such as SO_2 can remain trapped in Io's weak gravity field. Energized by sunlight, lighter molecules zip around Io at such high speeds that they ultimately reach escape velocity and leave the moon forever: conceivably, great amounts of water vapor and carbon dioxide were exhaled from the mantle of Io early in the moon's history, but these lighter molecules promptly reached the upper realms of the thin atmosphere, where their constituent atoms were dissociated by sunlight and bled out to space.

Although most of the heavy sulfur and sulfur dioxide stays prisoner of Io's gravity field, a small fraction does manage to escape: this trickle of ionized S and SO_2 has amassed into a tenuous trail in the wake of Io, funneled into a torus around Jupiter by the planet's strong magnetic field.

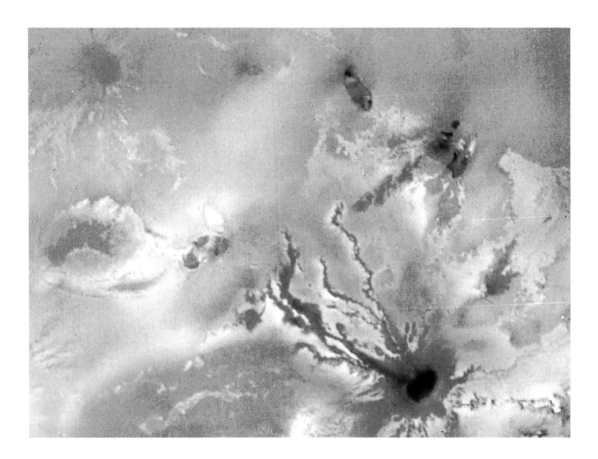

The color of sulfur

Sulfur not only dominates the atmosphere and cosmic environment of Io: it is also thought to rule the surface, judging from the colorful flows that streak the landscape, which have been tentatively ascribed to sulfur or sulfur-bearing magmas. Sulfur is an easily fusible element, which makes it a prime candidate for volcanic activity at low temperatures: sulfur melts between 383 K (110 °C) and 392 K (119 °C), depending on its allotrope (the element has a choice of a dozen possible molecular arrangements).

Sulfur is a complex element that displays a vast range of colors in its liquid and solid states. It is yellow around its melting point, but turns orange if the melt is heated; grades to red around 450 K; and to dark brown at higher temperatures. These diverse colors might sometimes be 'frozen in place': for instance, a very hot sulfur magma that erupts dark brown may initially stay brown if chilled, although with time it will revert to reds, oranges, and yellows as it adjusts to a cold environment.

Io's wonderful palette of colors could then be taken to reflect the range of chilled allotropes of sulfur, and early studies of flows on the Voyager imagery attempted to fit this model. The flows of Ra Patera, for instance, were seen to change from dark brown at the vent to red downstream before fanning out into broad orange lobes, a pattern that would fit the cooling trend of sulfur. This simple interpretation was quickly challenged, however, on the grounds that deep bodies of molten sulfur such as flows and ponds should cool slowly rather than quench, and have time to revert to yellow rather than hold on to 'hotter' colors. Instabilities within flows would also churn the melt and complicate the picture, as would impurities such as potassium and sodium salts in the melt that would shift the color scale (and such salts are known to be present in the moon's environment).

Sulfurous magmas

Despite the difficulties in interpreting the observed patterns, sulfur does seem to be the dominant color-making element on Io, and two extreme models compete to explain the volcanic process by which sulfur is spread onto the surface: dominantly sulfur melts, and silicate melts containing accessory sulfur.

In the all-sulfur model, a layer of molten sulfur several kilometers thick is capped by a thin crust of solid sulfur: 'windows' in the crust channel eruptions of molten sulfur to the surface. In the competing model, volcanism on Io is silicate-based, like on all terrestrial planets, and flows are basaltic. Sulfur would then simply exist in concentrations large enough to tint the flows, although it could build some secondary flows of its own.

Be it a pure magma or simply a colorful secondary element, sulfur plays an important role in the volcanism of Io, and this has led planetary geologists to review the role of sulfur on Earth for comparative purposes, and study the rare cases of sulfur flows on our planet.

The majority of terrestrial sulfur sank out of view in the early days of our planet's history, following iron in its catastrophic migration to form the core (sulfur follows iron in most of its physico-chemical adventures: it is said for this reason to be *siderophile*). Whatever sulfur is left in the mantle today readily enters melting phases, and can make its way back to the surface during eruptions: being a very volatile element, sulfur exsolves from the melt as soon as the drop in pressure permits, and joins the gas phase, jetting out of craters and coating vents and fissures with precipitates of yellow sulfur crystals, and colorful salts. These deposits are minor, and do not ordinarily give rise to sulfur flows. However, sulfur flows can develop under unusual circumstances, namely when high heat flow across volcanic ground melts the sulfur crystals and salts deposited in fissures, and drives the low-temperature melt to the surface.

Few flows of the kind have been described on Earth: one was caused by the eruption of Japan's Siretoko volcano in 1936, which so heated the volcano flanks that a chocolate-brown flow of sulfur oozed out of a fissure, and ran down the slope to form a tongue over 1 km long and 20 m wide: cooling slowly, it turned yellow as it solidified.

A smaller sulfur flow was created by the 1950 eruption of Mauna Loa in Hawaii: the feature was revisited by planetary geologists in the early eighties, in the wake of the Voyager mission and the discovery of sulfur-like flows on Io. On the Hawaiian site as well, the source of sulfur was traced to fumarolic deposits, remobilized by a

FIG. 9.6. The viscosity of molten sulfur as a function of temperature. Sulfur melts around 110 °C but instead of becoming consistently more fluid with increasing temperature, its viscosity shoots up between 150 °C and 200 °C before resuming a declining 'normal' trend. The various colors of the sulfur melt as a function of temperature are also indicated. Credit: From David Rothery, *The Satellites of the Outer Planets*, Clarendon Press, Oxford, 1992.

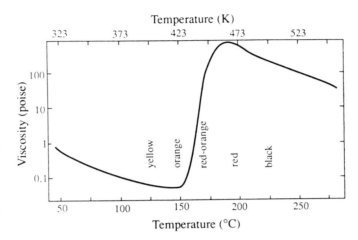

conventional eruption, and extruded as a short flow 27 m long, 14 m wide, and less than a meter thick. The flow exhibits features similar to those of basalt flows, albeit in scaled-down form, including tiny channels with levees; lobes extending away from the channel axis, less than a meter long; and even a miniature sulfur tunnel, 30 cm wide.

A perilous field trip

On Io, flows are up to 700 km long: if they are indeed made of sulfur, the scale is very different than on Earth, but physically reasonable. Sulfur is very fluid when molten, and channels and tubes would allow long run-off distances if the eruptive rate were large enough.

The cooling behavior of sulfur also follows a very original trend. Most molten substances – like silicate magmas – are fluid at high temperatures, and become more viscous as they cool and approach their solidification temperature. Sulfur behaves differently (see fig. 9.6). As a hot, dark brown melt (above 520 K), it flows fluidly, and increases in viscosity as the temperature drops. Around 470 K – in its red phase – sulfur magma becomes quite pasty. So far, nothing abnormal. But as the temperature keeps dropping, the viscosity trend suddenly reverses: viscosity plummets as sulfur enters its orange phase, down to a record low of 7 centipoise at 420 K – a viscosity ten thousand times lower than when it was hotter and red! The anomalous fluidity lasts over a few degrees of cooling, before the melt resumes a normal trend: as its enters its yellow phase, sulfur sees its viscosity shoot up again with falling temperature, coming to a halt and completely solidifying between 392 K and 383 K.

Imagine, therefore, the surprise of a volcanologist on Io, milling around an erupting sulfur flow. As it pours down the slope, the chocolate-brown river starts to cool, and slows its progression, inviting the unsuspecting geologist to move in for a sample. But then, as the lava keeps cooling and turns orange, it suddenly speeds up into a gushing river, sending our volcanologist running for safety!

One can also wonder what would happen each time hot sulfur runs over an older sulfur deposit. Solid sulfur doesn't need much energy to melt anew, and the overrun deposit could remobilize, perhaps in a more fluid state than the hotter overlying flow, depending on their respective temperatures. The result could truly be a mess, with different generation flows intricately mixed together.

Sulfur flows are not the only danger awaiting a manned expedition on Io. Explosive vents, spraying jets of frosty sulfur dioxide far and wide are another spectacular form of activity. The giant plumes pictured by Voyager are the most obvious eruptions, but there are also scores of brightly-colored blotches scattered across the surface (see fig. 9.5), which testify to recent blow-outs.

Crustal structure of Io

Many frost patches are found along scarps and cliffs, for there is substantial relief on Io in the form of plateaus and mountain blocks: apparently, rigid, rocky lithosphere does not lie far below the sulfurous surface, and emerges in many places. Indeed, sulfur alone cannot sustain such outstanding relief: even when solid, its plasticity is such that it would deform and flatten rapidly. Massifs 100 km wide and 10 000 m tall, as observed near Io's south pole must therefore be silicate bedrock, or at least contain appreciable amounts of silicates or other 'strengthening' compounds. This observation supports the theory that 'traditional' silicate volcanism takes place under the sulfurous envelope of Io, the crust swelling over thermal anomalies and breaking to form tilted blocks and scarps in tectonic zones.

Io might then have a crustal structure made up of both silicate and sulfur layers. At depth, tidal heating would keep the moon's silicate mantle partially molten, up to 30 km from the surface. From that level upward, solidified silicates would constitute the base of the crust, surging here and there to account for the relief at the surface. In the hollows of this silicate fabric, molten sulfur, seeped up over billions of years of igneous activity, would pool to form magma lakes several tens or hundreds of meters thick, crusted over by a surface shell of solid sulfur—itself conceivably up to one thousand meters thick.

Volcanism on Io might then operate in the following fashion: basaltic lavas would erupt through the fractures of the silicate crust onto the floor of the molten sulfur bodies—much in the same way that abyssal eruptions spread onto our ocean floors underwater on Earth. These high-temperature silicate melts would in turn stimulate the sulfur pool above them, triggering convection currents which would occasionally rip the uppermost sulfur crust, with flows of molten sulfur spreading at the surface, and jets of gaseous sulfur blowing sky high. The 300 km tall Pele plume could be such a jet of sulfur vapor, blowing out of an overheated sulfur pool.

The sandwich structure, with its lower base of silicate crust, spread with molten sulfur, and capped with an icing of solid sulfur, is but a model, and a simple one at that. The truth is probably much more complex, with a multitude of alternating layers of silicate and sulfur.

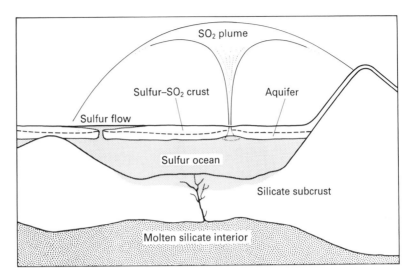

FIG. 9.7. Volcanic processes on Io. In this simplified model, a partially molten silicate mantle is overlain by silicate crust. In the topographic lows, molten sulfur has collected in deep pools, crusted over by solid sulfur. Liquid sulfur dioxide (SO_2) circulates through this upper, porous substrate, vaporizing over hot spots to form geyser-like plumes. At depth, silicate magma episodically rises to flow on the floor of the sulfur pools, and at the surface sulfur flows likewise emerge from fissures and vents. Credit: From B. Smith *et al.*, *Nature*, **280**, 741, © 1979 Macmillan Magazines Limited.

The geysers of Io

Another level of complexity is added if we consider the top crust to be somewhat porous: it might then act as a 'nappe', channeling any fluids that might exist at the reigning temperatures and pressures – notably sulfur dioxide.

Sulfur dioxide becomes liquid at temperatures and pressures corresponding to burial depths of about 1000 m. From that level downwards, SO_2 could circulate as a phreatic fluid in the crust, across temperature and pressure gradients. In the vicinity of molten bodies of sulfur and other hot spots, the compound would vaporize and blow out at the surface in geyser-like fashion (see fig. 9.7).

Calculations show that liquid SO_2 flowing in a porous substrate at a depth of 1500 m and meeting a pocket of hot molten sulfur would expand into a liquid/vapor emulsion, and – if given the chance to blow out through a vent – accelerate to exit velocities around 500 m/s. This figure matches the velocities derived for most plumes on Io (apart from Pele), lending credence to the SO_2 fountaining model.

One would then have two types of eruptions on Io: relatively low temperature SO_2 fountaining responsible for the Prometheus, Volund, Amirani, Maui, Marduk and Masubi plumes. On the other hand, Pele would be a much hotter, energetic type of eruption, perhaps involving molten silicates subliming a pool of sulfur as described in the previous section. Finally, the Loki plume might fall somewhere in between these two end models.

A comparison has been drawn between Io's SO_2 plumes and terrestrial geysers. Geysers are found in volcanic regions of the Earth, where high temperature gradients run close to the surface, and groundwater expands to vapor, spraying a liquid/vapor emulsion skyward. Iceland, New Zealand, and Yellowstone in the United States are among the Earth's principal geyser areas. On our planet, the high atmospheric pressure limits the amount of gas decompression that takes place at the vent – and thus the 'nozzle' exit velocity – so that the most energetic geysers (such as

Yellowtone's Old Faithful) do not spray their water plume higher than 100 meters. By comparison, calculations show that if Old Faithful were transplanted to Io, and fired up on the small, airless moon, the brutal decompression of hot water in the near vacuum and lower gravity would propel the plume of vapor and ice crystals to an altitude of 35 kilometers!

The number of frosty, sulfurous patches on the surface testifies to the extent of geyser-like activity that takes place on Io. The high concentration of white patches at the base of cliffs and scarps might point to lateral blasts, where liquid SO_2 running through the porous crust rushes out of the cliff face (the way water springs out of rock faces on Earth), spraying an emulsion of gas and frosty particles over the surrounding terrain.

Duration of eruptions

How long do eruptions last on Io? Scientists were given an extraordinary opportunity to tackle the question, as the Voyager program involved twin spacecraft: Voyager 2 arrived at Jupiter in July of 1979, four months after its predecessor. Initially, Voyager 2 was not programmed to observe Io, because its trajectory through the Jovian system was ill-adapted to an observation sequence of the inner moon. But in view of the exciting Voyager 1 discoveries, the second spacecraft was reprogrammed to direct its cameras toward Io during the flyby.

Despite a greater distance, Voyager 2 was able to locate seven out of the nine plumes observed by Voyager 1. One was out of view of the cameras, while another – and not the least – was missing altogether: Pele, the largest eruption on Io, had shut down, suggesting that it was indeed a special case, with a different origin than the other plumes. Despite the shut-down, Pele's ejecta could be spotted on the ground around the dormant vent: interestingly, the deposits were now symmetric, rather than heart-shaped, as in the Voyager 1 encounter. This meant that sometime between March and July, Pele had 'cleared its throat' and blasted away whatever deflected its gases, finishing its eruption in full symmetry.

Another volcano which evolved significantly between flybys was Loki, which showed different ground markings under its double plume. Loki's eruption was even seen to flare up when Voyager 2 was a few days away from closest encounter, reaching heights of about 300 km before settling back to 150 km. This time, it was the eastern vent which was the focus of most of the activity at Loki, spraying a much larger plume than the western vent.

Markings on the ground also showed substantial changes in Loki Patera, where the northern two thirds of the caldera was brighter and the southern part darker than during the Voyager 1 encounter. But the greatest surprise came from Aten and Surt calderas, which showed up much darker in the Voyager 2 imagery, as if they had recently been resurfaced by lava flows or lava lakes. Even more outstanding, the 50 km-wide calderas were surrounded by ejecta deposits 1400 km in diameter, as if two giant eruptions of the Pele type had occured between the Voyager encounters.

This would bring the number of high-energy plumes to three (Pele, Aten and

Surt), all apparently short lived. Pele's eruption lasted at least four days – the interval between the first and last plume image taken by Voyager 1 – but was extinct four months later when Voyager 2 swung by. As for the eruptions of Aten and Surt, they both lasted less than four months since they erupted between the flybys, leaving a trace without getting caught in the act. It appears from these three events, taken as a group, that Pele-type eruptions occur most probably over periods of days to weeks.

On the other hand, all eight events characterized by low and less energetic plumes (Prometheus and the like) were observed by both probes, as if they had never turned off between the flybys. For this pattern to occur, eruption durations of Prometheus-type plumes are more on the order of years: probabilities call for an average mean lifetime in excess of 32 months.

This two-classed behavior – short versus long eruptions – is again best explained if Pele and Prometheus-type plumes are driven by different mechanisms. In the Pele model, underground molten silicates vaporize sulfur to high temperatures and vent velocities, a behavior that could not last for very long because a reservoir of sulfur boiling off in an eruption would have trouble being replenished by sulfur circulation through the crust at the same rate: upon reservoir depletion, the eruption would stop. On the other hand, Prometheus-type 'geysers' of liquid SO_2 could go on erupting for much longer because liquid SO_2 happens to be very fluid (more fluid than water) and could drain from great distances across the aquifer to feed the erupting vents.

Hot spots on Io

Besides their imaging capability in the visible and ultraviolet, the Voyager spacecraft were equipped with infrared spectrometers (IRIS), and collected thermal profiles of Io's disk. This thermal data proved crucial, both in estimating Io's heat flow, and in refining the two-class model of plume eruptions.

At Jupiter's great distance from the Sun, Io's surface temperature should have read little more than 120 K ($-150\,°C$), but the spectrometric data showed temperature spikes much in excess of that figure, in the form of ten major hot spots. Eight could be modeled to represent temperature maxima between 221 K and 395 K, and were found to match the position of calderas on the visible imagery, including Babbar Patera and the Amirani/Maui plume region. A ninth, intense feature coincided with the Loki caldera/plume complex, and was modeled as representing a hot source at 450 K, surrounded by a wide area radiating at 245 K. Finally, the hottest spot of all was centered on Pele, with a calculated temperature maximum of 654 K.

Again, it appears from the thermal data that Pele is a very different feature than the smaller SO_2 plumes, with a thermal signature high above the melting point of sulfur. Actually, the 654 K temperature is about that needed to vesiculate boiling sulfur to the point of pyroclastic disruption and high-speed ejection out of a vent. One interesting corollary is that for sulfur to be so much hotter than its melting temperature requires some extraordinary heating source: one obvious mechanism would be for molten silicates to come in contact with a sulfur reservoir, and so the

654 K signature could be an indirect proof of the existence of molten silicates at depth.

If Pele is the hottest hot spot on Io, Loki is by far the most powerful. Although most of the feature is probably no hotter than 245 K (surrounding a smaller source at 450 K), it extends over such a large area (200 km × 200 km) that its power output is estimated at 12^{12} watts, more than all the other hot spots of Io put together. To account for this intense radiation one can speculate that the large, dark floor of the Loki caldera – in the shape of a horseshoe – is a pool of molten sulfur, although crusted over for the most part. Again, as with Pele, the Loki sulfur reservoir could be kept at high temperatures by erupting silicates on its floor – like a boiling pot over a stove.

Besides Pele, Loki, and the other eight major hot spots revealed by the Voyager 1 survey, careful reprocessing of the infrared data has recently brought out a dozen smaller features, much less energetic but still dwarfing anything seen on Earth (each radiates ten to one hundred times more heat than the Hawaiian or Réunion hot spots). These secondary thermal anomalies are associated with two Prometheus-type plumes (Marduk and Volund), and also with areas that could well be warm lava fields, erupted several years prior to the Voyager 1 encounter and in the process of cooling: one such region is found west of Pele, and comprises a dozen calderas with visible lava fields, including the spectacular Ra and Kibero Patera.

Since the Voyager flybys, more infrared observations of Io have been conducted from Earth, using powerful telescopes. The small apparent size of Io makes it difficult to identify specific regions, although this lack of spatial resolution can be compensated by a clever form of temporal resolution: each time Io passes through Jupiter's shadow, the fluctuation of its infrared signal (as its disk comes in and out of eclipse) reflects the successive locations and relative intensities of hot spot 'contributors' to the signal.

There are also sporadic infrared events on Io, observed in ground-based surveys to 'flash on' for hours or days: these are probably linked to major eruptions. One such energy burst was spotted in the spring of 1979 between the Voyager flybys, and most probably recorded the large Surt eruption that was missed by the spacecraft but left deposits on the ground for Voyager 2 to see. Its radiating wavelength of 5 micrometers corresponds to source temperatures of 500 to 600 K. One event observed in 1986 was even hotter: radiating at 3.8 micrometers, it implied temperatures in excess of 900 K, which would be the first strong evidence of silicate magma erupting at the surface of Io.

These ground-based surveys have shown Io's hot spots to be radiating just as much today as they were in 1979 during the Voyager flybys – especially Loki, which continues to display a prodigious, albeit irregular output.

Besides allowing the location and study of specific eruptions, the infrared data makes it possible to derive the global heat flux for Io. Planet-wide, the heat flux is calculated to be at least 1.5 watt per square meter, a figure twenty times larger than the average heat flux on Earth (0.075 W/m²). Revised estimates, taking into account the smaller hot spots recently identified, even reach 3 W/m². If this latter figure is correct, then Io's heat flux exceeds the theoretical maximum that tidal heating can provide in a stable, steady-state configuration (2 W/m²). What this might mean is that Io swings between periods of low tidal disturbance and periods of greater-than-average tidal distortion and heating – the kind that Io would be traversing today. Calculations based on tidal resonances between Io, Europa and Ganymede indeed indicate a periodic solution, where high heat-flow episodes would last 20 to 30 million years, followed by quiescent epochs about 100 million years long. Alternatively, the fluctuations could simply be random.

Crustal recycling on Io

To get another perspective of Io's activity, we can estimate the rate of resurfacing of the surface by lava flows and aerosols dispersed by the erupting plumes.

With respect to pyroclastic aerosols, global rates of deposition depend on the eruptive flux at each vent, and on the size of the particles: rough estimates indicate that the surface of Io is coated at a rate of a few millimeters per year, and perhaps as much as a few centimeters. This should be taken as a global average, since deposits will be thicker in the vicinity of vents, and thinner in other areas. As a first approximation, however, it appears that calderas and dormant vents (surrounded by sulfurous frosts) are equally distributed across Io, and that the particle coating should be fairly uniform – although curiously, the vents that were active in 1979 were nearly all grouped close to the equator, between $30°$ N and $30°$ S.

A few millimeters of global deposition per year is an impressive figure: at the lower end of the scale (1 mm/year), a layer of sulfur and sulfur dioxide flakes would build up to a thickness of one kilometer over the entire surface in only a million years; and at the higher end of the estimate (1 cm/year), up to ten kilometers of pyroclastic deposits could be laid down per million years.

Lava flows would contribute just as much matter, if not more to the surface. The eruptive rates that best match the infrared thermal data yield a thickness of one to ten meters of fresh lava per thousand years, spread over the entire surface of Io (in other words, again one millimeter to one centimeter per year, or one to ten kilometers per million years).

These estimates are validated by the near-to-complete absence of impact craters on Io. Considering the flux of asteroids and comets in the neighborhood of Jupiter, one can calculate the rate of lava and pyroclastic deposition needed to erase impact scars on Io, as fast as they form. The answer turns out again to be a minimum of one millimeter per year.

If we consider as a working hypothesis that the volcanic resurfacing on Io is 3 mm/year, then it would take a mere 10 million years for a layer 30 km thick to build up – the assumed crustal thickness of Io. Since the thickness of the lithosphere should remain constant, however (if the moon is in thermal equilibrium), then we must consider the 10 million year figure to be the 'turnaround' time of the crust – i.e. the period over which the crust is entirely recycled by volcanic activity. Over a 4 billion year history, this would amount to 400 cycles of complete overturn.

The arrival of the Galileo spacecraft at Jupiter in December of 1995, sixteen years after Voyager, will undoubtedly answer many questions, and raise new issues. Will the same plumes rise high above Io's horizon, or new ones carry the torch? Will there be fresh, new lava fields and calderas? Despite the blockage of its high-gain antenna – crippling the transmission of data back to Earth – Galileo should shed new light on the activity of extraordinary little Io, as well as on the nature and history of the other Jovian moons.

Cryovolcanism

Io is an exception among the satellites of the Giant Planet, by virtue of its tremendous heat flux, its perennial eruptions, and its dense silicate body. Most other moons around Jupiter, Saturn, Uranus, and Neptune, are lower density objects, made up of water ice, other volatiles and silicates in various proportions. Churned by heat pulses at various times in their history (depending on their size and tidal influence), most of these moons display extensional graben and flows of viscous ice at their surface – and in one special case operating geysers of low-temperature volatiles (Neptune's moon Triton). This form of activity and landform shaping on the icy moons of the Solar System has been dubbed *cryovolcanism* – volcanism of the cold.

Can one still speak of volcanism when the magma is water slush? Certainly, since water is just as much a mineral as are silicates and metal oxides: it has a solid, crystalline state (ice), with low temperature phase changes (melting, sublimation). The changes are so familiar to us and so widespread on Earth that we hardly think of them as 'volcanic' manifestations. Who would say of a melting glacier that it is 'erupting', and that a river of melt water rushing down a valley is 'lava'?

Ice is certainly a very special 'mineral'. Besides the fact that it melts at a low temperature (273 K), and that it is a remarkable chemical solvent when liquid, water has the particularity of increasing in density (shrinking in volume) when it melts. This is contrary to most other elements and compounds in nature, which instead expand and become buoyant when they melt – a driving force for the rise of silicate and sulfur magmas to the surface of planets. Because of its reversed density behavior, melt water would not be expected to rise through its icy matrix to the surface of a moon and participate in cryovolcanic activity.

However, the ices in the outer Solar System are impure, mixed with other frozen volatiles (namely ammonia) and silicate rock: these 'dirty ices' behave differently than pure ice, and experience classic phase changes with expansion during melt and buoyancy of the liquid phase. These are the ices that fuel cryovolcanism and give birth to flows and icy sprays at the surface – the exsolution of volatile impurities providing a supplementary ascensional force in the form of expanding gaseous phases.

The other moons of Jupiter

Starting in the Jupiter system, the moon closest to Io – Europa – brought us our first glimpse of cryovolcanic terrain. Europa is slightly smaller than Io (3100 km), and almost as dense (3 g/cm³). Circulating on a wider orbit, it is responsible, as we previously saw, for the tidal 'pumping' that drives Io's volcanism.

Conversely, Io exerts a symmetric tug on Europa, also distorting its orbit. For

Europa, the consequences are much milder than for Io, since it is farther away from Jupiter, and the forced eccentricity imposed on its orbit takes place within a much weaker gravitational field. Calculations show that tidal heating should be fifteen times weaker for Europa than for Io, and indeed the moon does not show any contemporary, spectacular activity. However, the absence of impact craters any greater than 20 km in size indicates that Europa's surface is young – 100 million years or less, according to some estimates.

The surface of Europa is nearly flat and consists of water ice, criss-crossed by a complex network of graben – tears in the crust hundreds of kilometers long and up to 100 km in width. Radiogenic heating early in the moon's history might have kept the ice partially molten, until the heat ran out and the crust froze and expanded, cracking into the array of faults observed at the surface. Depending on the extent of tidal heating, 'ice slush' might still upwell to this day, oozing through the faults to form the mottled regions on Europa (patches of dirty ice), and the bright strips in the bottom of some grabens (clean fresh ice). The bright surface of the plains might likewise be a cover of recent ice flows or pyroclastic frost deposits.

With virtually no tidal distortion and much less radiogenic heating than Europa (because of a lesser concentration of silicates), icy moons Ganymede and Callisto ran out of radiogenic energy much earlier in their history: ice melting and flowing was probably limited to the first billion years or so of their evolution. Their old, icy crusts are pock-marked with impact scars, especially Callisto which is globally saturated with craters. Ganymede displays a mix of old, dark terrains, and younger, brighter terrains. These latter patches show up as interlaced bands of grooves and ridges, that must have been the theater of most the icy upwellings early in Ganymede's history.

The moons of Saturn

After the Jupiter flyby, Voyager 1 and 2 continued their trek across the outer Solar System, reaching Saturn in November of 1980. Around the giant, ringed planet, the probes imaged the enormous moon Titan (about as large as Mars but, to the disappointment of planetary geologists, covered by a haze of ammonia and methane blocking out all surface detail); and icy worlds Iapetus and Rhea (1500 km in diameter); Dione and Tethys (1100 km); Mimas and Enceladus (500 km); as well as a host of smaller-sized satellites.

The moons of Saturn have densities between 1.1 and 1.4 g/cm³, pointing to ice contents in excess of 60%. Ices are indeed abundant at the low temperatures that prevail so far out from the Sun (80 K) – not only water ice but also other frozen volatiles like ammonia and methane. Ices in the Saturn system and beyond are therefore predominantly compound ices, such as ammonia hydrate ($NH_3.2H_2O$) and water *clathrates* hosting trace amounts of other volatiles in their crystalline lattice. These clathrates melt around 170 K ($-100\,°C$) and thus need little energy to mobilize. Contrary to pure water, they rise readily to the surface upon melting since they are less dense in their liquid than in their solid form.

On Saturn's moons, the Voyager cameras revealed smooth landscapes attributed to these compound ices, often torn and rifted by graben. On Dione, the plains are criss-crossed by long trenches; and on Tethys, a gigantic rift tears through the cratered crust, probably a tectonic readjustment to a giant impact on the opposite hemisphere.

But it is Enceladus that displays the youngest surface, with no impact craters and such fresh-looking ridges and plains that the diminutive moon is believed to be cryovolcanically active at present. The plains are so bright that they must have recently been resurfaced, either by ice flows or by explosive eruptions spreading frosty particles far and wide. Possible proof of this ongoing activity is provided by one of Saturn's rings – the 'E' ring – which coincides with the orbit of Enceladus as if it were a trail of frost injected in space by the moon's explosive eruptions. Since dynamic calculations show that the ring should disperse in less than 10 000 years, the last cryovolcanic eruptions on Enceladus would have to be younger than this – a hint of contemporary activity.

What energy source is responsible for Enceladus' volcanic longevity? As was the case with Io and Europa in the Jovian system, the answer seems again to be tidal heating. Enceladus feels the wrenching tug of Saturn when it wanders off its circular orbit, which it periodically does under the influence of the outer, larger moon Dione. Calculations show, however, that in its current orbital configuration, Enceladus should only experience tidal heating on the order of 100 million watts – one or two orders of magnitude less than the power needed to melt its mantle of ice. To account for Enceladus' hyperactivity, a stronger orbital eccentricity is needed to pump up the tidal heating. Such a forced eccentricity might be provided by yet another moon of Saturn – diminutive Janus – joining forces with Dione to periodically place a stronger tug on Enceladus.

The moons of Uranus

After the successful Jupiter and Saturn encounters, only Voyager 2 was on the proper trajectory to pursue its trek across the Solar System, reaching Uranus in January of 1986 and imaging a new family of icy moons: Titania and Oberon (1700 km in diameter); Umbriel and Ariel (1200 km); and smallest, innermost Miranda (470 km).

The moons of Uranus are just as varied as the systems of Jupiter and Saturn, each body showing a distinctive character: all seem to have suffered, at one time or another, from crustal splitting and cryovolcanism. In some cases, lobes of ice were imaged in such detail that educated guesses could be made as to flow rates and viscosities. From spectral analyses, the water ices appeared rich in hydrated ammonia, with traces of nitrogen, methane, and carbonaceous dust.

Many tectonic features on these satellites are reminiscent of rift structures on Earth. Titania displays canyons up to 50 km wide and thousands of meters deep, with one large chasma comparable in morphology to the Horn Graben of the North Sea.

FIG. 9.8. A Voyager 2 view of Ariel, one of the larger moons of Uranus (1300 km in diameter). Impact craters testify to an old, presently inactive surface but linear grooves and bright patches indicate that Ariel was active in the past with faulting and cryovolcanic eruptions. Credit: NASA/JPL (Voyager).

Ariel also displays many large rifts (see fig. 9.8), with viscous-looking ice flows running down the length of many graben.

The steeper flow fronts on Ariel are calculated to correspond to a viscosity of 10^4 to 10^8 poise, similar to that of a rhyolite flow on Earth, and this could be due to ammonia hydrates at 176 K, clogged with crystals. More fluid-looking flows could be crystal-free melts, with viscosities around 10^2 poise – closer to that of an andesite magma on Earth. Therefore the variety of flow morphologies on Ariel could be explained by varying degrees of crystal concentration in the 'slush'.

Eruption processes on Ariel can also be modeled from the aspect of ice flows on the graben floors. Two alternatives are proposed (see fig. 9.9): either ice oozed up all along the faults, spreading on both sides of a central fissure; or else the ice flows erupted from scattered vents along the graben floors, building ice tubes down their lengths. In favor of the latter, many graben flows on Ariel are reminiscent of the basalt tubes that snake down the course of many Arizona canyons.

The biggest surprise in the Uranus system came – again – from the innermost moon. Instead of showing a dead, cratered surface as its small size led to expect, diminutive Miranda (470 km in diameter) turned out to be torn by strange, concentric fault 'rings' – features that were first named *ovoids* and later *coronae*.

There are three large coronae on Miranda, with angular corners and a striated appearance. Two of the coronae – Arden and Elsinore – are somewhat rectangular, while the third feature – Inverness – is closer to a trapezium. All three are rung by fault scarps, with a slightly depressed, chaotic central arena. In the case of Elsinore and Inverness, this central part is slightly undulated, swells that could be caused by cooling, shrinking, and wrinkling of cryovolcanic flows, or alternately by the piling up of flows above fissures.

FIG. 9.9. Two models for the emplacement of ice flows in the grabens of Ariel. In model A, viscous ice upwells through an axial fissure running down the length of the graben and spreads laterally to both sides. In model B, ice upwells in a more fluid state through discrete chimneys to flow down the length of the graben, isolated by a rook of crusted-over ice, in lava-tube fashion. Lateral stretching (bottom) would then collapse the roof of the tube. Credit: From S. K. Croft and L. A. Soderblom, *Uranus*, p. 599, Editors J. T. Bergstrahl, E. D. Miner, and M. S. Matthews. © 1991 University of Arizona Press.

A. Axial eruption model

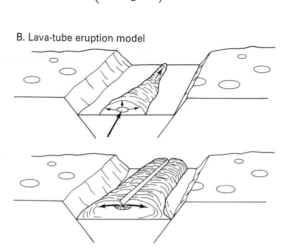

B. Lava-tube eruption model

To explain Miranda's coronae and cryovolcanic flows, two contrasting theories have been put forward. In the tidally-induced plume-rising model, Miranda's orbit is occasionally distorted through resonance with its neighbor Umbriel, and the moon submitted to a heating pulse within Uranus' strong gravitational field. The ensuing meltdown would have sent ice plumes rising to the surface, stretching the crust to form coronae, and feeding flows of viscous ice onto the surface.

The second, original model calls for a radically opposite mechanism. During its tumultuous youth, Miranda might have broken up as the result of collisions and reaccreted out of the scattered blocks: three late-accreting blocks would have plowed through the icy asthenosphere, sinking and setting up icy convection currents in their trail. The coronae would then be the icy scars of these extraordinary, engulfment impacts.

The geysers of Triton

In the wake of its successful Uranus encounter, Voyager 2 was redirected toward Neptune, last objective of its multi-planet mission. The probe reached the outer giant planet on August 25, 1989.

Neptune was known to have only one sizeable moon: Triton (2700 km in diameter). From Earth-based spectrometric analysis, Triton was thought to be covered by nitrogen ice with some methane, and Voyager confirmed the prediction, revealing a complex landscape varnished with nitrogen clathrates (including minor

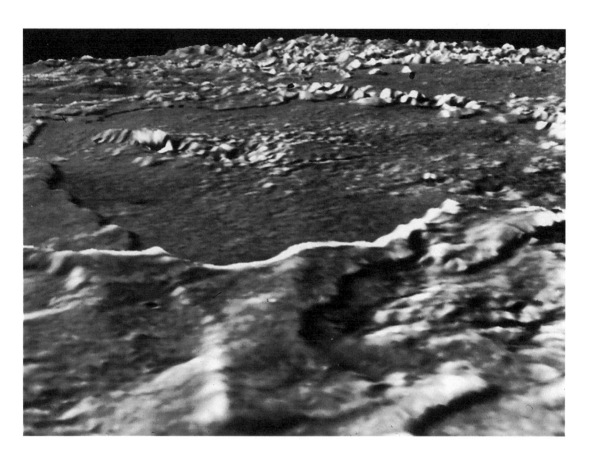

FIG. 9.10. Computer-generated perspective of an icy caldera on Triton, obtained from a Voyager 2 image (vertical exaggeration: 20 ×). The caldera floor is approximately 200 km in diameter, with a maximum relief of 1000 m. The caldera floor is flat and surfaced with ice flows: its central region (left of center in the image) is particularly rich in pits and flows. Credit: NASA/JPL (Voyager).

amounts of methane and carbon monoxide ice). Over 4 billion kilometers from the Sun, the faint light flux reaching Triton still managed to set off photochemical reactions in the ice, transforming methane into colorful polymers and giving Triton its pastel blend of blue, yellow, and peach tints, under a haze of wind-blown gas.

The small number and subdued appearance of impact craters on Triton indicated a substantial level of cryovolcanic activity, and indeed a multitude of small vents were spotted on the Voyager imagery, as well as smooth plains assumed to be outpourings of clathrate ices. Four large depressions up to 200 km across showed up in the smooth plains, looking like icy calderas, with pits and flows concentrated in the center (see fig. 9.10).

Older terrain on Triton appeared pock-marked by a dense array of subcircular cells, 5 to 25 km in diameter, traversed by sinuous ridges. Giving Triton's surface the appearance of a melon skin, this so-called cantaloup terrain covers roughly a third of Triton's observed surface, and is reminiscent of salt diapir provinces on Earth. One explanation is that the crustal ices of Triton were heated at some stage in the moon's history (perhaps when Triton was captured by Neptune), and warm ices of various compositions (nitrogen, hydrated ammonia, methane, carbon monoxide) erupted onto the surface. Upon cooling, some of the deeper ices ended up more buoyant than

FIG. 9.11. Artist's view of an erupting geyser on Triton. The plume of expanding nitrogen gas surges high above the icy surface of the moon. A second plume is visible on the horizon, with giant Neptune hanging in the airless sky. Credit: Painting by William K. Hartmann and Ron Miller.

the overlying ones, and rose through them as pancake-shaped diapirs, to create the cellular patchwork at the surface.

During its speedy flyby, Voyager 2 took a dozen images of Triton from a variety of angles, and overlapping images were paired to create stereoscopic views of the surface. While pairing-up two images, Voyager team member Larry Soderblom discovered to his great surprise that two horizontal plumes stood out above the surface, stretching westward from what looked like erupting, dark columns 8 km high. Geysers on Triton! Comparing images taken 45 minutes apart, the wind-blown plumes showed a dynamic growth, doubling in length from 80 to over 150 km. Near the easternmost event, at least one other diffuse plume was located, probably the subdued trace of an eruption that had only recently shut down. By combing and contrast-stretching the rest of the imagery, another half-dozen veils were highlighted above the moon's limb.

Geysers thus appear to be a widespread occurence on Triton. Regarding the eruption process, a crucial clue was provided by the geographical distribution of the plumes and streaks: all were bunched in the austral latitudes between 40° S and 60° S – the region pointing toward the distant Sun. Triton's geysers therefore appear to be solar driven, although temperatures at the surface are so low (around 38 K) that even nitrogen ice would not be expected to melt under normal conditions. Imaginative heating mechanisms were proposed, including greenhouse effects occuring in the translucent ices. Indeed, Triton might be the first world that we encounter where eruptions are driven by solar thermal heating!

Solar heating on Triton would work along the same principles as our solar-heating panels on Earth: in these, a transparent surface (usually glass) covers a layer of air or water, coated at the bottom with a dark substrate. Sunlight streaking through the glass is absorbed by the dark substrate, which radiates the energy back at longer wavelengths (infrared). But since the glass cover is opaque to infrared, the energy stays trapped within the pannel, raising the heat of its fluid (air or water).

Triton's surface could likewise be a natural solar panel: a superficial layer of frozen, transparent nitrogen (a couple of meters thick) would act as the transparent cover and medium: visible solar radiation penetrating the surface would be converted to infrared by dark, carbon-like impurities in the ice (playing the role of the dark substrate) and the resulting infrared radiation would remain trapped. With a temperature rise of only a few degrees, a fraction of the nitrogen ice would then begin to vaporize and circulate through fractures in the 'panel', until reaching a vent and decompressing to blow out as a plume.

So ends our tour of the icy moons of the giant planets – the wonderful geyser display on Triton capping a rich review of cryovolcanic features across four families of satellites. In the closing chapter, we summarize the observations made across the Solar System, to draw a global view of planetary volcanism.

10 Volcanism: a planetary perspective

Planetary exploration

As the fourth decade of space exploration comes to a close, the human species has successfully toured the Solar System, and surveyed the main geological processes at work on the rocky planets.

The sixties set the bases of interplanetary travel, and were the golden years of lunar exploration. In the seventies, concurrently with the close of the Apollo program, attention shifted to our neighboring planets Mars and Venus, unveiled by automated orbital and landing missions. While missions continued to Venus, the eighties were marked by the Voyager missions to the outer giant planets, and their rocky and icy moons. The nineties are capping our reconnaissance of the Solar System with a climactic radar survey of Venus (Magellan), and a closer look at the smaller bodies populating interplanetary space – asteroids and comets.

Throughout these pioneering years of planetary geology, the deluge of scientific data has grown in step with the ever-increasing sophistication of remote-sensing equipment, and a nearly exponential rise of radio transmission rates. The Magellan mission to Venus is a good example: the probe collected a greater volume of data in three years around Venus (1990–93), than had all missions to the planets combined over the thirty preceding years.

On the eve of the twenty-first century, a technological mutation is taking place with the development of small, economic spacecraft to complete the armada of heavy planetary 'clippers'. The Clementine lunar mission of 1994 ushered in this new trend, marking the US return to the Moon after twenty years of neglect. Meanwhile, the last great missions of the century include a long-awaited return to Mars (the American Mars-Surveyor and Pathfinder missions, and the Russian Mars 96 and 98), and the first orbital missions to the giant planets – Galileo, due to arrive at Jupiter in December of 1995; and Cassini, expected to reach Saturn by the year 2000.

Mercury, the forgotten one

Because of the energy needed to reach Mercury, and the time needed to reach Pluto, the closest and farthest planets to the Sun were long ignored. It is only around the year 2000 that a miniaturized probe will finally lift off for distant Pluto, an icy planet smaller than Mars that holds captive a moon the size of our own (Chiron).

Much closer to the Earth, spinning fast and close to the Sun, Mercury was studied and photographed by the US probe Mariner 10 in 1973–74, over the course of three speedy flybys. The constraints of celestial mechanics were such that always the same half of the planet's surface was viewed by the skirting probe at each encounter. The images broadcast back to Earth showed Mercury to have a relatively old surface, riddled with impact craters. The planet boasts one giant impact basin (Caloris), that

FIG. 10.1. Mercury is battered by impact craters like the Moon, but lacks the vast basalt mare that are so typical of our satellite. Smooth plains buckled by wrinkle ridges (top half of image) might be the only impression of volcanic flows emplaced in the early history of the planet. Alternatively, these units could be sheets of ejecta driven from nearby impact basins. Credit: NASA/JPL.

lacks the lava fill so typical of its lunar equivalents. Barely larger than our Moon, Mercury apparently lived just as short a thermal history, with even fewer volcanic outpours.

One major surprise was the global set of compression folds buckling the mercurian plains, as if the entire planet had shrunk in the distant past and wrinkled like a desiccated apple – perhaps as a result of global cooling, or phase changes in the iron nucleus densifying and contracting the planet.

With respect to volcanism, evidence is scarce on Mercury, limited to patches of smooth plains contrasting with the older cratered terrain: these might well be lava flows (see fig. 10.1). An alternative hypothesis is that the deposits are ejecta sheets from large impact craters.

This lack of obvious volcanism on Mercury can be explained if we consider the small size and the structural make-up of the planet: Mercury has only a thin mantle draping the bulky iron nucleus, and such a thin mantle can only stock a limited amount of heat-releasing radioactive atoms. Besides its limited internal heating, Mercury is given to rapid cooling: its small size gives the planet a more balanced area to volume ratio than Mars, Venus and the Earth, and conduction alone is sufficient to evacuate the heat flow, without resort to convection, melting, and volcanism – except perhaps in the early days of its history, when the radiogenic output was at its peak, and strong magnetic currents also reached out from the young Sun.

Volcanism on asteroids

There is one last class of planetary bodies in the Solar System that deserves attention from a magmatic and volcanic perspective: asteroids and comets, which are primeval blocks of metal, silicates, ice and other frozen volatiles.

Evidence from meteorites (chips of asteroid and comet matter fallen to Earth) indicates that many asteroids underwent heating and igneous differentiation in the early days of the Solar System, despite their small size. A whole class of meteorites – the *eucrites* – are basaltic in nature, and represent 5 to 15% partial

FIG. 10.2. Artist's view of a probe landed on a comet. Asteroids and comets have become the target of a number of interplanetary missions at the close of the twentieth century. Large asteroids are suspected of having undergone magmatic differentiation and volcanic eruptions early in their history. Farther out in the Solar System, comets are ice-rich and undifferentiated: their activity is limited to the sublimation of volatiles when they swing in close to the Sun, a form of cryovolcanism displayed here by a 'geyser' on the horizon to the right. Credit: Painting by William K. Hartmann.

melting of their parental bodies, whereas other meteorite groups are partial melt residues, such as the coarse-grained *lodranites*, and the ultramafic *ureilites*.

Several competing processes are proposed to account for asteroidal meltdown and petrogenesis, including radiogenic heating, impact heating, and electromagnetic induction heating. Impact heating is not the most favored, because meteoritic samples from asteroids are highly differentiated – a process that would not occur in impact melts, except if they were very large and stratified through the action of an important gravitational field.

Induction heating, on the other hand, was certainly operative in the early days of the Solar System, when it was caused by the strong magnetic field of the young Sun. It was a unique process in that it heated the entire bodies exposed to the field, independently of depth or elemental composition. Rising melt, in particular, would keep receiving heat all the way up to the surface, in contrast to conventional magma plumes that receive all their heat at depth, and progressively lose it during ascent.

Conventional, radiogenic heating is of course the other prime mechanism to explain asteroid differentiation. Elements of short half-life 'flared' vigorously in the early days of the Solar System, heating and melting even small bodies that contained them. Fast-decaying elements that were abundant at the time included Al-26 and Fe-60, with half-lives of under one million years.

What did eruptions look like on rocky bodies a few tens to a few hundred kilometers across? The first magmas rising to the surface were likely to be volatile-rich (estimates call for several thousand parts per million of CO, N, and Cl), and explosive eruptions would have occured in the nearly complete vacuum of asteroidal and cometary environments. Projections might easily have reached the escape velocities of most of these bodies for volatile contents as low as a few hundred parts per million.

On an asteroid 100 km in size for instance, 1000 ppm of volatiles is sufficient to propel the disrupted magma to vent velocities in excess of 100 m/s, sending erupted fragments out into solar orbit. In the early days of the Solar System, most fragments would likely have spiraled into the Sun, and disappeared from the planetary record. This would explain why there are relatively few basaltic specimens found among meteorites.

For the larger asteroids like Vesta (280 km diameter), escape velocities are more important (close to 500 m/s) and for ascending melts to reach such velocities at the vent would require at least 3% of volatiles in weight, which appears unrealistic. Therefore, basalt fragments erupted on such bodies would tend to fall back onto the asteroid's surface, and the few basaltic eucrites drifting in space today are rather due to impact ejection of surface rocks from their parent bodies than to eruptive blow-outs.

One particular class of lavas chipped off the surface of asteroids by impacts, and collected on Earth as meteorites, are the *pallasites* which are rich in minerals troilite (iron sulfide) and olivine. These highly differentiated rocks point to very large ratios of partial melting in their parental bodies (up to 50%), which is to say that magma oceans were at work on these predominantly molten bodies. Such strong heating and meltdown of early asteroids was more likely caused by magnetic induction than by radiogenic heating.

Magma oceans on asteroids certainly started off as turbulent bodies, driven by convection cells where gravity was sufficiently strong: we can imagine a 'scum' of chilled material developing and breaking continuously at the surface, and a layer of dunite cumulates settling at the bottom. Such differentiation would be limited to the larger asteroids.

The low gravitational field of the smaller bodies would limit any complex differentiation of the melt, since crystals of different densities would migrate too slowly to organize in layers before cooling chilled the whole batch to a halt. One particularity of magma oceans on small bodies is indeed that they would cool very fast once the heat supply shut down, because of their relatively low volume and thermal inertia: on small asteroids, shallow magma oceans would have solidified in a matter of decades.

The planets: a comparative study

As we saw in Chapter 1, planets and other solid bodies of the Solar System have three ways of evacuating their internal heat: simple conduction of the heat flow through a solid lithosphere up to the surface; partial melting of layers at depth, with molten material rising in plume fashion to the surface and physically carrying along the heat (hot spot mechanism); and – in a variation of the latter – foundering of cold and dense crustal slabs into the mantle, balanced by the convective rise of hot mantle (plate tectonics model). Volcanism at the surface occurs in the two latter cases.

Small planetary bodies are usually content with conduction alone, because their advantageous area-to-volume ratio ensures that the bulk of their heat is evacuated as fast as, or faster than it is generated at depth. Larger planets, on the other hand, suffer

FIG. 10.3. Ternary diagram showing the relative importance of the three main modes of heat evacuation within planetary bodies: non-volcanic conduction, volcanic hot spots, and plate recycling. This classification highlights the predominance of conduction for small bodies (the Moon, Mercury, and even Mars), which figure close to the conduction apex. Larger Venus is farther down the line toward the hot spot apex since the latter process represents a more sizeable fraction of its heat output. Exceptional Io evacuates most of its tidal-generated heat through hot spot plumes. Finally the Earth is close to the plate recycling apex, the process which removes the bulk of its heat (conduction is second, and hot spots a distant third). Credit: Based on a diagram by J. W. Head, Brown University.

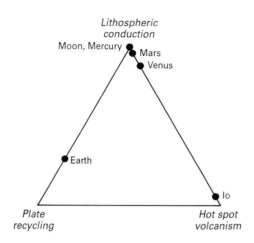

from an unbalanced area-to-volume ratio, heat up faster than they can cool by conduction alone, and call into play either or both of the melt-transport mechanisms (hot spot plumes and plate tectonics).

To represent the relative importance of cooling mechanisms in each planet, a ternary diagram can be used with conduction, hot spot, and plate tectonics figured at each apex (see fig. 10.3).

In such a diagram, small bodies like the Moon and Mercury are represented close to the conduction apex since they evacuate nearly all of their heat by this process (the 'hot spot' volcanic plains of the Moon represent a negligible contribution to the overall heat budget). A larger planet like Mars is pictured farther away from the conduction apex, since hot spot volcanism on the red planet contributed slightly more cooling over time than in the case of the Moon and Mercury: perhaps as much as 2% of the total (conduction accounting for the other 98%).

At another corner of the diagram, Io stands out as an exceptional body: despite its small size, Jupiter's moon has to rely heavily on hot spot volcanism to evacuate the tremendous amount of heat generated by tidal disturbance.

The largest terrestrial planets – Earth and Venus – occupy more balanced positions on the diagram, blending several processes in their cooling regime. Our planet evacuates a very small fraction of its heat through hot spots (probably around 5%) and relies mostly on conduction and plate tectonics. Conduction within plates does away with 30% or so of the planet's heat (at a rate of 75 milliwatts per square meter at the surface), whereas plate tectonics account for over 60% if we include the high heat flow at the mid-ocean ridges (enhanced by hydrothermal circulation), mid-ocean ridge volcanism, and heat transmission to cold slabs in subduction zones.

Finally, Venus remains the most enigmatic planet to date. We have little indication of its global heat flow but it is believed that conduction evacuates at least 75% of it, and that melting and volcanism accounts for less than a quarter. Indeed Venus displays many hot spot landforms (volcanic rises, giant shields, novae and coronae), but their contribution to planetary cooling looks relatively minor over time.

One question that remains open, however, is the possibility that heat build-up in the Venus mantle led in the past to resurfacing events on a much grander scale.

The time factor

The relative unimportance of magmatic heat evacuation on the Moon and Mars might come as a surprise in view of the striking volcanic landforms that adorn these bodies. But however spectacular the lunar mare or the giant shields of Mars may be, their relative importance is scaled down to perspective if we consider that they represent most everything volcanic that ever occured at their surface. Over four and a half billion years of history, they turn out to be minor contributors to the global cooling.

As an example, Olympus Mons may well be over 500 km in width and 27 000 m tall, it was built up over hundreds of millions of years – perhaps over a billion – and its contribution to planetary cooling was very modest. It is estimated that for the red planet as a whole over the last billion years, lava flows did not exceed one or two hectares per year, two orders of magnitude less than on Earth.

In the case of the Moon, activity is weaker yet, especially in the recent past since there does not seem to have been a single volcanic eruption in the last billion years. Even the older mare account for a very small proportion of the heat flow, since these relatively thin bodies of lava were emplaced over millions of years, their flows separated by millennia of stillness.

It is therefore important to distinguish between sporadic eruptions – spectacular as they may be – and global averages calculated over aeons of geological time.

An activity scale

If we now wish to grasp the total extent of volcanic activity in the life of each planet, and compare figures from planet to planet, we can turn to the volume of igneous crust generated over each planet's lifetime, divided by that planet's crustal volume: such ratios are convenient in order to compare activities on different-sized bodies.

The Moon, for starters, has a 60 km-thick crust generated in its early history (4.5 to 4.3 billion years ago). Since then, volcanic lavas poured over 20% of the lunar surface, with cumulate thicknesses ranging from several hundred meters to several kilometers. If we take one kilometer as an average, and were to spread the lavas over the entire lunar surface for representation purposes, we would end up with a 200 m-thick layer. Compared to the 60 km-thick crust, the contribution of the maria can then be expressed as an addition of only 0.3% of new matter (200 m divided by 60 km) in the course of 4 billion years.

If we now turn to Mars, estimates of lava extent and thicknesses are more speculative, but the consensus is that all lava fields and shields add up to about 1% of the martian crustal volume. For small, 'normal' planets, it thus appears that volcanism accounts for very little crustal build-up indeed, after the initial formation of the crust.

The addition of igneous products to the primeval crust soars dramatically when we consider larger planets like Venus and Earth, where trapped internal heat fueled volcanism at high rates over long periods, and renewed the crust to a much larger extent.

For Venus, figures are speculative but volcanic output over time might be as high as two to three times the initial crustal volume – which we express as 200 to 300% crustal contribution. This does not mean that the crustal thickness of Venus has doubled or tripled over time: as new lavas were erupted and loaded at the top, the crust was depressed under the extra weight, and its bottom layers pushed down to high pressures where phase changes and densification might well have peeled them off, to sink and melt back into the mantle. The addition of new lavas atop the Venus crust would then have been balanced by the 'delamination' of older lavas at the bottom, with the crust keeping a constant thickness as a net result. Rather than crustal growth, we should then speak of a 300% 'crustal recycling' of Venus.

If Venus recycled its crust two to three times over its history, the Earth did much better: the oceanic lithosphere is thought to have entirely disappeared in subduction zones (while regenerated elsewhere as fresh lava and intrusives) as much as a dozen times over geological history – a crustal renewal factor of over 1000%!

This would constitute the record of igneous activity in the Solar System if it were not for exceptional little Io: at the observed rate of plume eruptions and sulfur flows on Jupiter's satellite, the making of a layer 30 km thick (the assumed size of the crust) might take less than a dozen million years. This would then mean that Io has renewed its crust by volcanic processes close to 500 times during its history – a recycling rate of 50 000% on our comparative scale.

From less than 1% crustal renewal (the Moon) to 1000% (the Earth) and 50 000% (Io), the importance of volcanism in planetary processes is most variable indeed.

Global asymmetries

Volcanism is rarely a homogeneous process on any planet, and often affects certain areas preferentially to others. This geographical asymmetry is obvious in the case of the Moon, where the near side basins received most of the volcanic activity (the maria) while the far side stood relatively inert. This asymmetry, as we noted in Chapter 4, is probably due to the displacement of the Moon's center of mass toward the near side, bringing the top of the mantle there closer to the surface.

Mars also displays a global asymmetry, nearly as marked as in the lunar case: volcanoes and lava plains are concentrated in the northern hemisphere, namely in the boreal plains and the Tharsis and Elysium regions, while the southern hemisphere remains thick-crusted and inactive (except for the Hellas impact basin surrounded by paterae volcanoes).

The Earth is no exception. Two thirds of its surface have foundered to harbor oceanic volcanism, while the remaining third became sealed off by thick continental crust, letting only vigorous hot spots breach the lithosphere up to the surface. Apart from this oceanic/continental dichotomy, the Earth also displays an asymmetry in the arrangement of its deep mantle hot spots, which are not uniformly distributed.

FIG. 10.4. Sif Mons, a 300 km-wide volcano on Venus with a lava flow in the foreground. The shield towers 2000 m above the plains (relief is greatly exaggerated in this computer-simulated 3-D view). Sif Mons is located in Eistla Regio (21° N, 352° E), a broad volcanic rise probably due to a deep mantle hot spot. Volcanic features on Venus are concentrated in two broad areas known as BAT (Beta-Atla-Themis) and ATT (Alpha-Tellus-Tethis). Eistla Regio and Sif Mons are on the western edge of the latter. Credit: NASA/JPL (Magellan).

Two geographical families are prominent: one main family bunches under Africa and the Atlantic Ocean; and a more diffuse set of hot spots lingers under eastern Asia and the Pacific rim.

A two-fold distribution of hot spots is also inferred for Venus: volcanic features are found in higher than average concentrations in one large area known as BAT (Beta-Atla-Themis regions), as well as in a more diffuse area known as ATT (Alpha-Tellus-Tethis regions). The land between these areas is, on the contrary, relatively depleted in volcanic rises, shields and other igneous features.

If we turn to Io – the most active of planetary bodies – two apparent asymmetries can be noted. The first asymmetry is latitudinal: most erupting plumes were observed in the low latitudes – between 30° N and 30° S. Only plume number 8, and the eruptions of Surt and Aten (that took place between the two Voyager flybys) were higher up in the mid-latitudes (45°N to 48°S), but no eruption was observed closer to the poles. This concentration of plumes in the low latitudes is surprising, especially since calderas and vents in general are evenly distributed at all latitudes.

The second pattern on Io has to do with heat flow, and this time the asymmetry is longitudinal: most hot spots are concentrated in the leading, Jupiter-facing part of the moon, a longitudinal band centered on the 300° W meridian (240°–360° W, with a peak heat flow at 310° W). Interestingly, the three Pele-type, high-energy eruptions (Pele, Surt and Aten) also fall in this sector, which is noticeably redder than the rest of the planet – probably because of the abundance of sulfur pyroclasts. Although it might be tidal in origin, this longitudinal asymmetry still remains unexplained.

Convection patterns

Patterns in the distribution of volcanic features often reflect the geometry of underlying convection currents. Convection currents are influenced in turn by the location of heat sources within a planet. Simulations show that if the heat sources are concentrated in the mantle (as are radiogenic atoms), then convection currents tend to spread out as multiple small plumes. Inversely, if a substantial proportion of a planet's heat is generated at greater depth (as residual heat or phase change energy in the core), then deeper-rooted, larger plumes emerge.

The Earth is a case in point: today, the vast majority of its heat is radiogenic and comes from the mantle (the core only accounts for an estimated 10% of the global heat flow). Theoretical calculations predict that this type of heat distribution should lead to many small plumes, and this is indeed what is observed: the Earth boasts forty to fifty, relatively small hot spots.

In the case of Venus, thermal models call for even less heat issued by the core, and correspondingly few large and deep-rooted plumes. The distributions of topographic swells and gravity anomalies on Venus seem to confirm the presence of half a dozen large hot spots – the volcanic rises of the Atla and Beta class – whereas the coronae landforms would be the expression of shallower, smaller hot spots generated in the upper mantle.

The heat budget of planet Mars is modeled to have been more core-dependent than for the Earth and Venus: predictions call for ten or so very large hot spots rooted deeply in the red planet. This time, however, observations do not match the model: there only seem to be two hot spot provinces on Mars – those associated with the Tharsis and Elysium uplifts. Perhaps Mars started off with a greater number of plumes, as the model would have it, but blended them for some unknown reason, so that buoyant mantle material ended up rising under only two areas. Another possibility is that the heat contribution from the martian core has simply been overestimated.

Volcano families

Plumes and convection currents regulate the distribution, dimensions and lifetime of volcanic provinces on the surface of planets. On a finer scale, the shape and structure of individual volcanic features are controlled by regional factors, such as the tectonic

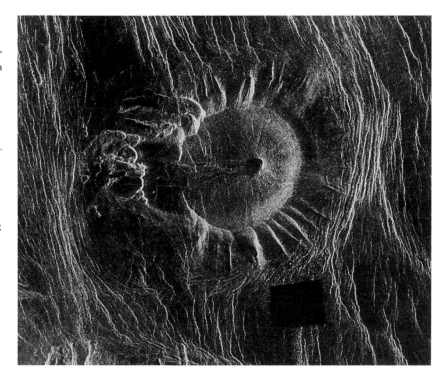

FIG. 10.5. A 'tick' on Venus (18° S, 5° E in Eistla Regio). This 65 km-wide feature belongs to an original class of volcanic domes on our sister planet, characterized by a relatively flat, concave summit, and fluted flanks. Lava flows originate in a dark central pit, and breach the summit's western rim. Regional north–northwest trending lineaments (graben) are deflected around the feature, especially to the east. Ticks are one of several classes of original volcanoes on Venus that seem to have no obvious counterparts on Earth. Credit: NASA/JPL (Magellan).

setting that affects the dimension and orientation of fractures; and by the environmental characteristics of each planet: the strength of the gravitational field and the atmospheric pressure.

Before the space age, volcanic features as they were defined on Earth fell into five broad categories: featureless lava plateaus and plains (traps and the like); gently-sloping shields; steeper lava shields and ash cones; rounded domes of viscous lava; and large, fault-rung calderas, associated with ignimbrites and other ash-fall deposits.

The exploration of other planets extended the size-range of existing volcano classes: giant shields were discovered on Mars, two orders of magnitude more voluminous than their equivalents on Earth; large 'pancake domes' on Venus – 50 km wide as opposed to kilometer-sized domes on Earth; and gigantic calderas on Mars, Venus, and Io.

Our neighboring planets also presented us with original, unexpected features: large paterae on Mars – in the shape of flattened, overturned saucers; and enigmatic coronae on Venus with subdued relief and concentric fracture rings, as well as arachnoids, novae, and ticks (see fig. 10.5).

In the light of comparative analysis, even the Earth possesses a unique class of volcanoes, apparently shared with no other planet: resurgent calderas, surrounded by vast ignimbrite fields. Resurgent calderas are characterized by ring fractures and 'trap door' faults, and are driven by cycles of domical bulging, catastrophic eruption, and collapse. However the fact that these features have not yet been discovered on other planets does not mean that they don't exist there: silicic deposits are difficult to

FIG. 10.6. Schematic diagram of major volcanic landforms on Earth, arranged by quantity and quality of lava. Individual lava flows and cinder cones result from small quantities of magma, whereas giant shield volcanoes and ignimbrite calderas are due to large volumes of mafic and silicic magmas respectively. (Original classes of extraterrestrial volcanoes are not represented here). Credit: From *Moon and Planets*, by William K. Hartmann, Wadsworth Publishing Company.

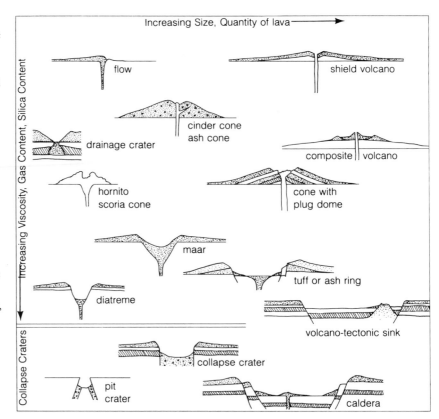

identify on any planet, starting with the Earth as we saw in chapter 2, and more work is needed before we can confirm the absence of ignimbrite resurgent domes outside Earth.

From the confrontation of all interplanetary data, we can now attempt to sort out the general rules of volcanism from their planet-dependent particularities.

The shaping of a volcano

The shaping of volcanoes throughout the Solar System is governed by a number of parameters including tectonic setting and gravitational field; rheology and chemistry of the magma (including temperature and volatile content); and the discharge rate during eruptions (see fig. 10.6).

One of the most important parameters is the discharge rate – the magma flow rate at the vent. High discharge rates of mafic lava lead to flood basalts – vast expanses of lava with little to no relief. Heat is so efficiently retained in these voluminous outpours that viscosity is kept very low and lavas spread over large areas. Eruptions of this kind created the traps and great underwater plateaus of the Earth; the basaltic mare of the Moon; and extensive volcanic plains on Mars and Venus.

Where magma is rich in volatiles – be it from a water, carbon, or sulfur-rich environment – high discharge eruptions can turn explosive above certain threshold levels, and the result will again be low relief: base surges and convective columns will spread ash over large areas, especially on small planets where gravity is low and atmospheres thin or non-existent.

Lower discharge rates during volcanic eruptions promote steeper forms of relief. In the case of lava flows, a lower discharge rate means less volume and thermal inertia: flows cool faster, cover less range, and pile up closer to the vent. The result is a range of forms, from sloping shields to steeper cones and domes. Where volatiles are involved, low discharge rates are also associated with relief – stratocones built of both lava and ash; and cones and domes of nuée ardente deposits (except again where gravity is weak and pyroclasts are spread far and wide).

Volcano height

While discharge rate is mainly responsible for the size and slope of volcanic landforms, the maximum height that a volcano can reach is quite another matter.

The maximum height reached by a fluid is ruled by the laws of hydrostatics, which involve the density difference between the ascending magma and encasing wall rock: the greater the density difference, the greater the ascensional force of the magma according to Archimedes' law. Equally important is the depth of the source where the magma originates: the deeper the source, the higher the level reached by the lava above the surface.

The equations indicate that for a typical density difference of 5% between magma and encasing rock, the maximum theoretical height that a volcano can reach is one twentieth the depth of the source. If the density difference is 10%, the maximum height is one tenth the depth of the source.

The reason mid-ocean volcanoes on Earth build to relatively low heights above the sea floor (a few hundred meters) is thus mainly because their magma is derived from shallow melt zones, at depths around 5 km. Similarly, volcanoes are low on Venus (shields average 1500 m) probably because the crust is thin, and underlying magma sources are relatively close to the surface, at depths on the order of 15 to 30 km.

On the opposite end of the scale, Hawaiian volcanoes reach heights of 10 000 m above the sea floor because they derive their magma from great depths in the Earth's mantle (around 100 km). Even more extreme are Olympus Mons and the giant Tharsis Montes of Mars, that owe their record heights of over 25 000 m to very deep magma sources (250 km if the magma/wall rock density difference is 10%).

One should not forget of course that hydrostatic equations only indicate the maximum theoretical height that a volcano can reach. Volcanoes grow only if other conditions are satisfied – namely if discharge rates are low enough to promote the building of relief, as we just saw. Mare magmas on the Moon surged from depths of hundreds of kilometers but because the discharge rate was so high, they spread out far from the vents rather than built shields.

The gravity factor

With respect to discharge rates, volcano heights, and other parameters that shape volcanic landforms, gravity does not seem to play a fundamental role. In the equations of discharge and flow rates, gravity only shows up as a simple factor, on equal rank with magma density, density difference with encasing rock, viscosity, elastic limit, and fissure length – the whole lot overshadowed by fissure width which is raised to the square.

When it comes to lava flow away from the vent, gravity plays an equally minor role. The maximum runoff distance of a flow is again governed by the eruptive rate, where gravity is of little influence. Where gravity does play something of a role is that it allows levees of cooled rock to build up higher on the margins of lava channels in low gravity environments, allowing them to carry thicker, more insulated flows farther downstream. All in all, a weak gravitational field does allow lava to flow farther than a stronger field does, but the effect is minor: all other conditions being equal, lava on the Moon will flow 1.7 times farther than on Earth, although their gravity forces differ by a factor of 6.

Where gravity does exert a strong influence on volcanic landforms is in the case of pyroclastic projections. During a volatile-rich eruption, scoria are projected along parabolic trajectories shaped by the gravitational field. Where gravity is strong (as on the Earth and Venus), projected particles will fall back close to the eruptive vent, building up steep spatter cones and cinder cones. On the other hand, where the gravity field is low (as on the Moon and Io), trajectories of bombs, scoria, and ash will be stretched, and the ejecta distributed over greater areas. Pyroclastic volcanoes on the Moon and Io are likely to be flattened disks of fall out.

The pressure factor

Gravity also has a bearing on the type of environment in which eruptions will take place. Small, low gravity planets do not retain much of an atmosphere, and eruptions at their surface occur in near vacuum. Where volatiles are present in the magma, disturbance and pyroclastic activity are promoted – as exemplified by the glass beads brought back by the Apollo astronauts, that attest to past episodes of lava fountaining on the Moon; and the erupting vents on Io, that spray umbrellas of sulfur high over the landscape.

Larger planet Mars presents an interesting case on our scale of planetary environments. On the one hand, the red planet has a very tenuous atmosphere: the pressure today only reaches 600 pascals at the surface – a mere two percent of the pressure on Earth. Volatile expansion is nearly as complete as in vacuum, and where volatiles are not lacking, results in explosive behavior. Indeed, the ancient 'Dandelion' paterae of Mars and younger features like Alba Patera and Hecates Tholus show evidence of pyroclastic eruptions – aprons of soft ash, fluted and gouged by erosion.

On the other hand, Mars did not always have a thin atmosphere. The red planet traversed warmer epochs when volatiles were massively released from the subsurface, and pressures rose to tens of thousands of pascals, perhaps up to half the pressure we experience on Earth today. These were times on Mars when the atmosphere interacted strongly with eruptions: volatile expansion and disturbance were limited by the higher pressures, but the interaction of groundwater and ice with rising columns of ash complicated the picture, as did the engulfment of atmospheric gases in plinian and pelean eruptions – energizing base surges, and lofting ash convectively to great heights.

As we turn to larger planets with thicker atmospheres, pyroclastic eruptions are possible only for high volatile contents in the magma, that are able to overcome burdening pressures at the vent. The Earth succeeds in fostering explosive activity because of the high concentrations of water and carbon dioxide in its mantle and crust. A range of explosive activity is observed, from hawaiian lava fountaining and strombolian projections to plinian plumes and pelean base surges. On Venus, where pressures of millions of pascals bear down on the vents, only the highest volatile concentrations might manage to disturb the magma explosively.

Atmospheric pressure also has an influence on the size of ejecta propelled in the course of an explosive eruption: in high pressure environments, erupting gases stay dense as they blow out of a vent, and are capable of lofting heavy particles into the air. On low pressure planets, erupting gases expand so much at the vent that they lose their ability to entrain heavy particles. On Mars, for instance, where the atmospheric pressure is two hundred times weaker than on Earth, the maximum size of entrained particles is two hundred times less: martian ejecta will be micrometric to millimetric in general, centimetric at best.

On Earth, the denser gases are capable of lofting blocks several meters in size: bombs up to one hundred tons in weight have been found ten kilometers away from their source vents. On Venus, where the atmosphere is eighty to ninety times denser than on Earth, explosive eruptions – if they occur at all – would have a lifting power capable of lofting blocks ten to twenty meters large, and weighing thousands of tons.

Volcanoes and atmospheres

Over the course of planetary evolution, volcanic eruptions have had a great influence on the development of atmospheres and the shaping of climates.

Atmospheres on terrestrial planets have largely grown from the degasing of volatile-rich mantles: these gases are predominantly water vapor and carbon oxides; and to a lesser extent argon, nitrogen, sulfur compounds, chlorine and other halogens. Changes in these volcanic atmospheres have occured over time through gravitational grading of the gases – the lighter ones escaping to space; and through chemical cycles, as on Earth where carbon dioxide has for the most part dissolved in ocean water and entered carbonate minerals. Biological activity, where present, also modifies the gas line-up: oxygen and methane on Earth have grown out of photosynthesis, respiration, and other forms of metabolism of living organisms.

We can get a feeling for the gaseous contribution of volcanoes to planetary

FIG. 10.7. Ash-laden volcanic plume, extending out to sea from an erupting volcano on the Kamchatka peninsula, as viewed from the Space Shuttle (STS-68 mission, 1994). Other snow-capped stratocones can be seen left of the plume, along the coast. Volcanoes contribute large amounts of volatiles to the Earth's atmosphere – principally water vapor and carbon dioxide, and to a lesser extent sulfur dioxide, chlorine, fluorine and argon. Credit: NASA, courtesy of Debra Dodds, Johnson Space Center.

atmospheres by monitoring eruptions on Earth – although it is sometimes difficult to distinguish between juvenile volatiles issued for the first time from the deep mantle, and meteoric volatiles already exsolved in the crust and atmosphere, that are simply recycled in eruptions. Isotopic ratios, where available, can often be used to make the distinction.

Different types of magma carry different types of volatiles in varying amounts: on average, silicic magmas have higher concentrations of chlorine, water vapor, and carbon dioxide than basaltic magmas, but the latter are much richer in sulfur.

As for the amounts released, direct evaluations can now be made by analysis of the gas phase at the vent during an eruption, or by remote-sensing from planes and satellites. The Paricutin eruption of 1943, as an example, discharged 18 000 tons of water vapor per day during the growth of its small cone – water vapor being the principal gas phase of terrestrial eruptions. Mount Etna, apart from water vapor and other gases, exhales an estimated 35 000 tons of carbon dioxide per day, and has done so for millennia – the high figure being probably due to limestone country rock decomposed by the magma.

The El Chichon 1982 eruption is also of particular interest because its sulfur yield to the atmosphere was directly measured by airborne sampling and satellite surveys, and turned out to be phenomenal: about 6 million tons of sulfur dioxide in a month – 200 000 tons a day. Here, the high figures could be caused by the dissociation of sulfur-rich sediments at depth (anhydrite), so that a volcano's geological setting is often as crucial as the composition and yield of its juvenile magma.

Volcanoes and climate

Beyond the fact that they are responsible for building up planetary atmospheres, volcanic eruptions have a noticeable effect on the climate. Condensed droplets, ice crystals and dust particles create cloud formations that can reflect a large fraction of the solar radiation back to space, or alternatively trap a fraction of the incoming radiation at specific levels of the atmosphere, heating those layers and cooling others. Infrared-absorbing gases like water vapor and carbon dioxide can also generate greenhouse effects, by which the infrared heat radiated by the planet is trapped and drives temperatures up to high levels – Venus being a prime example.

The evolution of atmospheric temperatures and weather patterns is a complex combination of these varied and often opposite processes. The dual behavior of sulfur dioxide is a good example, which increases the temperature of the atmosphere by greenhouse effect as long as it remains in the gaseous state (SO_2), but decreases it when it converts to H_2SO_4 aerosols in the upper atmosphere and intercepts incoming radiation.

On Earth the effect of volcanic eruptions on climate was reported scientifically for the first time in 1784, when Benjamin Franklin, who was living in Europe at the time, made the connection between the unusually cold winter of that year and the monumental eruption of Laki in Iceland, that had spewed ash and gas into the atmosphere the previous summer and fall. Laki is believed to have injected less than a cubic kilometer of tephra but as much as 20 million tons of sulfur and 200 million tons of chlorine into the lower atmosphere. The 1784 winter was the coldest on record in two centuries, 5 °C below the long-term average. Because the dust and aerosols did not penetrate the stratosphere, they settled quickly and the climate recovered promptly.

The explosive eruptions of Tambora in 1815 and Krakatoa in 1883 were followed by equally noticeable and significantly longer cold spells. Tambora generated much ash (over 100 km^3) in its 1815 explosion, as well as an estimated 200 million tons of sulfate aerosols. This time the erupting column reached well into the stratosphere (probably up to an altitude of 50 km), and ice cores in Greenland indicate that the aerosols took four years to settle. Worldwide temperatures in 1816 dropped an average of 0.7 °C, and the climatic effect lingered for several years. In the 1883 Krakatoa eruption, 50 million tons of sulfuric aerosols reached the stratosphere: the solar radiation was observed to decrease by 10% around the globe and temperatures were depressed by an average 0.4 °C for three years.

What can be made of all this? First of all, it is believed that for an eruption to have a planetary-wide effect on climate, a large proportion of the ash and gases must reach heights of at least 10 km in the mid-latitudes (and 17 km in the tropics) in order to penetrate the jet stream and the stratosphere, stay suspended, and get spread out by the winds all around the globe. Secondly, it appears that sulfate and halogen aerosols have a greater effect on the climate than volcanic ash, because the tiny droplets reside longer in the atmosphere than ash, which tends to aggregate and fall out.

From this one would expect that explosive, plinian eruptions of silicic lava have more of an effect on climate than other types of eruptions, because they are observed to inject gases high into the stratosphere. Basaltic fire fountains, even though they release up to ten times more sulfur per unit mass than silicic eruptions, would be at a disadvantage. This could be true for average-sized eruptions, but less so when one considers the rare but phenomenal trap eruptions of the past.

Trap eruptions and volcanic winters

Trap eruptions consist of huge outpours of mafic flood lavas: because of the sheer volume of magma involved, the release of volatiles dwarfs any historic eruptions known to man. Fifteen million years ago, the Roza eruption of the Columbia River Plateau is estimated to have released 6000 to 10 000 million tons of aerosols – fifty times more than Tambora in 1815. Calculations show that if they were spread evenly over the globe, these aerosols would have intercepted over 99.99% of the sunlight worldwide, bringing the luminosity down to that of a moonlit night.

Again, how such a massive injection of aerosols might affect the climate depends on how high the aerosols are propelled and how long they stay suspended. If the injection takes place in the lower atmosphere, the aerosols would settle reasonably fast and have little effect. But there is reason to believe that such large eruptions, although non-explosive, can propel aerosols to great heights simply by virtue of the tremendous heat released. Laki's 1783 fire-fountain eruption gives us a hint of this thermal convection effect: following the eruption, a sulfurous haze was seen high in the Alps, very close to the stratosphere (but not quite making it). Convective columns over lava fountains were also reported to reach heights of at least 5 km over both the 1961 Askja eruption and the 1984 Mauna Loa eruption.

Convecting clouds over giant trap eruptions are expected to have risen much higher, and penetrated the stratosphere: instead of settling in a matter of months, the aerosols of eruptions like Roza might have stayed suspended for five to six years, with deep darkness and cooling disturbing the climate on a grand scale and setting off a chain of reactions in the biosphere.

Trap eruptions are even believed by some to be the cause of the major biological crises that have punctuated the history of evolution on Earth – the so-called mass extinctions. The Late Cretaceous crisis, that saw the brutal disappearance of the dinosaurs and of up to two thirds of all living species on Earth, appears to temporally coincide with the eruption of the Deccan Traps in India. However, the crisis is more convincingly tied to the giant impact of an asteroid or comet in the Yucatan (the Chicxulub crater). Proponents of the volcanic theory can still claim that the Deccan eruption contributed to the biological crisis – although to what extent is still a matter of debate. One theory even goes as far as to suggest that it was the Yucatan impact, by focusing seismic waves at its antipode, that triggered the melting of the upper mantle in the Deccan province and led to the trap eruption.

Other biological crises in the Earth's history can tentatively be paired up with periods of trap volcanism, starting with the great Permian/Triassic mass extinction of 250 million years ago, that coincides with the eruption of the Siberian traps. Coincidences of the kind should still be taken with a grain of salt, since other equally large eruptions are associated with no mass extinction whatsoever. But be it as it may, the study of catastrophic eruptions in the past and the modeling of 'volcanic winters' is of paramount importance in understanding the various stresses that disturb the biosphere, and could imperil the human species in eruptions to come.

The rains of Mars

Other planets help to illustrate the importance of volcanic gases in the shaping of atmospheres and climates – although of course they shed no further light on associated biological stresses since the Earth alone is graced with life.

On Mars, the Alba Patera volcano has served as a case study because of the morphological evidence of pyroclastic deposits on its flanks – pointing to volatile-rich eruptions. The ash cover is estimated on Alba to be more than 100 m deep over 500 000 km^2 – more than 50 000 km^3 in volume. If one assumes that the driving volatile was water vapor at a concentration of 1%, then the Alba eruptions could have contributed enough water vapor to form a condensed layer one centimeter deep over the entire planet. If, on the other hand, it was carbon dioxide that drove the plinian eruptions of Alba, calculations call for a total exhaust of 10^{15} kg of CO_2 – equivalent to 5% of the present atmosphere of Mars.

If a global estimate is made of all volatiles contributed by volcanic eruptions on Mars – explosive events and lava flows combined – and taking water vapor as the main volatile phase, then the martian shields and paterae could have outgassed close to 1.5 m in depth of water (once condensed) over the entire planet, and up to 10 m if the extensive volcanic plains are included in the calculations. In the early days of martian eruptions, volcanic gases might have built up to such volumes and pressures that shallow oceans could have pooled in the lows of the red planet, under higher temperatures due to the greenhouse effect. This hypothesis finds some justification in the sedimentary-looking features found in several basins of Mars, namely in Valles Marineris.

On Venus, on the other hand, the water vapor and carbon dioxide exhaled by volcanoes over the years were exposed to the higher solar influx of a much nearer Sun. The greenhouse effect was devastatingly efficient and rising temperatures barred water from ever entering the liquid phase: instead, the vapor was dissociated by sunlight over time, and bled out to space, leaving carbon dioxide as the main constituent of the atmosphere.

That the Earth avoided the sterile, ill fate of Venus and Mars is principally due to our planet's ideal location in the Solar System, far enough from the Sun to have avoided the runaway greenhouse effect that took control of Venus, yet not so far as to fall into a catastrophic freezing pattern, as did Mars when its declining volatiles were no longer sufficient to entertain the mild greenhouse effect that warmed it in its early history.

On Earth our volcanoes served us well, providing the ideal quantities and balance of volatiles for our planet to take advantage of its place in the Solar System. Water vapor condensed to form the oceans, carbon dioxide dissolved in the oceans before stirring up too much of a greenhouse effect, and with the evolution of life in such a temperate and fertile environment, creatures now exist that can marvel at the beauties of the world. Volcanoes especially will always hold a special place in our hearts – on Earth and beyond – as we continue to explore and learn more about the worlds that surround us.

Bibliography

The Earth (Chapters 1 and 2)

Books and compilations

Bardintzeff J.-M., *Volcanologie*, Masson, Paris, 1992.
Beatty J. K. and **Chaikin A.**, eds., *The New Solar System*, 1990.
Debelmas J. and **Mascle G.**, *Les Grandes Structures Géologiques*, Masson, Paris, 1993.
Francis P. W., *Volcanoes of the Central Andes*, Springer-Verlag, 1992.
Macdonald G. A., *Volcanoes*, Prentice Hall, Englewood (NJ), 1972.
Ozima M., *The Earth, Its Birth and Growth*, Cambridge University Press, 1981.
Simkin T., *et al.*, *Volcanoes of the World*, Smithsonian Institution, Huchinson Ross Publishing Company, USA, 1981.

Articles

Camus G., *et al.*, *Volcanologie de la Chaîne des Puys*, Aurillac, Clermont-Ferrand, 1982.
Coffin M. F. and **Edlholm O.**, Scratching the surface: estimating dimensions of large igneous provinces, *Geology*, **21**, 515–18, 1993.
Greeley R., The Snake River Plain, Idaho: representative of a new category of volcanism, *J. Geophys. Res.*, **87**, 2705–12, 1982.
Hekinian R., **Moore J. G.** and **Bryan W. B.**, Volcanic rocks and processes of the Mid-Atlantic Ridge rift valley, *Contrib. Mineral. Petrol.*, **58**, 83–110, 1976.
Rampino M. R. and **Caldeira K.**, Episodes of terrestrial geologic activity during the past 260 million years: a quantitative approach, *Celestial Mechanics and Dynamical Astronomy*, **54**, 143–59, 1992.
Simkin T., Terrestrial volcanism in space and time, *Annu. Rev. Earth Planet. Sci.*, **21**, 427–52, 1993.
Sleep N. H., Hotspot volcanism and mantle plumes, *Annu. Rev. Earth Planet. Sci.*, **20**, 19–43, 1992.
Smith D. K. and **Cann J. R.**, The role of seamount volcanism in crustal construction at the Mid-Atlantic Ridge, *J. Geophys. Res.*, **97**, 1645–58, 1992.
Smith R. L., Ash flow magmatism, *Geol. Soc. Am. Spec. Pap. 180*: 5–27, 1979.
Wilson L., *et al.*, Physical processes in volcanic eruptions, *Annu. Rev. Earth Planet. Sci.*, **15**, 73–95, 1987.

The Moon (Chapters 3 and 4)

Books and compilations

French B., *The Moon Book*, Penguin Books, 1977.
Hartmann W. K., *et al.*, eds., *Origin of the Moon*, Lunar and Planetary Institute, 1986.
Heiken G., **Vaniman D. T.**, and **French B. M.**, *Lunar Sourcebook: a User's Guide to the*

Moon, Cambridge University Press, 1991.

Moore P., *The Moon*, Mitchell Beazley Publishers, London, 1981.

Taylor S. R., *Lunar Science: A Post-Apollo View*, Pergamon Press Inc., New York, 1975.

Wilhelms D. E., *The Geologic History of the Moon*, U.S.G.S.P.P. 1348, Washington, 1987.

Wilhelms D. E., *To a Rocky Moon: A Geologist's History of Lunar Exploration*, University of Arizona Press, 1993.

Articles

Belton M. J. S., *et al.*, Galileo multispectal imaging of the north polar and eastern limb regions of the Moon, *Science*, **264**, 1112–15, 1994.

Greeley R., *et al.*, Galileo imaging observations of lunar maria and related deposits, *J. Geophys. Res.*, **98**, 17183–17205, 1993.

Head J. W., Lunar volcanism in space and time, *Rev. Geophys. Space Phys.*, **14**, 265–300, 1976.

Howard K. A. and **Head J. W.**, Regional geology of Hadley Rille, in *Apollo 15 Preliminary Science Report*, NASA SP-289, Washington D. C., 1972.

Ryder G., Apollo's gift: the Moon, *Astronomy*, **22**, 40–45, 1994.

Schmitt H. H., *et al.*, The Apollo-11 samples: introduction, *Proceedings of the Apollo 11 Lunar Science Conference*, **1**, 1–54.

Schmitt H. H., Evolution of the Moon: Apollo model, *American Mineralogist*, **76**, 773–84, 1991.

Schmitt H. H. and **Cernan E. A.**, A geological investigation of the Taurus-Littrow Valley, in *Apollo 17 Preliminary Science Report*, NASA SP-293, Washington D. C., 1973.

Spudis P. D. and **Greeley R.**, The formation of Hadley Rille and implications for the geology of the Apollo 15 region, *Lunar Planet. Sci.*, **XVIII**, 243–54, 1987.

Swann G. A., *et al.*, Preliminary geologic investigation of the Apollo 15 landing site, in *Apollo 15 Preliminary Science Report*, NASA SP-289, Washington D. C., 1972.

Wood J. A., The Moon, *Scientific American*, **233**, 93–102, 1975.

Mars (Chapters 5 and 6)

Books and compilations

Cattermole, P., Mars: *The Story of the Red Planet*, 1992.

Kieffer, H. H., *et al.*, eds., *Mars*, University of Arizona Press, Tucson and London, 1992.

Mutch T. A., *et al.*, *The Geology of Mars*, Princeton University Press, Princeton, 1976.

Articles

Baird A. K., *et al.*, The Viking X-ray fluorescence experiment: sampling strategies and laboratory simulations, *J. Geophys. Res.*, **82**, 4595–624, 1977.

Binder A. B., *et al.*, The geology of the Viking Lander 1 site, *J. Geophys. Res.*, **82**, 4439–4451, 1977.

Carr M. H., *et al.*, Some Martian volcanic features as viewed from the Viking Orbiters, *J. Geophys. Res.*, **82**, 3985–4015, 1977.

Clark B. C., *et al.*, The Viking X-ray fluorescence experiment: analytical methods and early results, *J. Geophys. Res.*, **82**, 4577–94, 1977.

Crumpler L. S., Aubele J. C. and Head J. W., Calderas on Mars: models of formation for the Arsia type, *LPSC* **XXII**, 269–70, 1991.

Francis P. W. and Wood C. A., Absence of silicic volcanism on Mars: implications for crustal composition and volatile abundance, *J. Geophys. Res.*, **87**, 9881–9, 1982.

Frey H. and Jarosewich M., Subkilometer Martian volcanoes: properties and possible terrestrial analogs, *J. Geophys. Res.*, **87**, 9867–79, 1982.

Mouginis-Mark P. J., Wilson L. and Head J. W., Explosive volcanism on Hecates Tholus, Mars: investigation of eruption conditions, *J. Geophys. Res.*, **87**, 9890–904, 1982.

Mouginis-Mark P. J., Wilson L. and Zimbleman J. R., Polygenic eruptions on Alba Patera, Mars, *Bulletin of Volcanology*, **50**, 361–79, 1988.

Mutch T. A., *et al.*, The geology of the Viking Lander 2 site, *J. Geophys. Res.*, **82**, 4452–67, 1977.

Plescia J. B. and Saunders R. S., Tectonic history of the Tharsis region, Mars, *J. Geophys. Res.*, **87**, 9775–91, 1982.

Reimers C. E. and Komar P. D., Evidence for explosive volcanic density currents on certain Martian volcanoes, *Icarus*, **39**, 88–110, 1979.

Scott D. H., Volcanoes and volcanic provinces: Martian western hemisphere, *J. Geophys. Res.*, **87**, 9839–51, 1982.

Sleep N. H., Martian plate tectonics, *J. Geophys. Res.*, **99**, 5639–55, 1994.

Solomon S. C. and Head J. W., Evolution of the Tharsis province of Mars: the importance of heterogeneous lithospheric thickness and volcanic construction, *J. Geophys. Res.*, **87**, 9755–74, 1982.

Tanaka K., Martian geologic 'revolutions': a tale of two processes, *Lunar Planet. Sci.*, **XXI**, 1237–8, L. P. I., 1990.

Toulmin P., *et al.*, Geochemical and mineralogical interpretation of the Viking inorganic chemical results, *J. Geophys. Res.*, **82**, 4625–34, 1977.

Watters T. R., Chadwick D. J. and Liu M. C., Distribution of strain in the floor of the Olympus Mons caldera, in *MEVTV Workshop on the evolution of magma bodies on Mars*, LPI Technical Report **90-04**, 293–4, 1990.

Wilson L. and Mouginis-Mark P. J., Volcanic input to the atmosphere from Alba Patera on Mars, *Nature*, **330**, 354–7, 1987.

Zimbleman J. R., Solomon S. C. and Sharpton V. L., The evolution of volcanism, tectonics, and volatiles on Mars: an overview of recent progress, *Lunar Planet. Sci.*, **21**, 613–26, 1991.

Zuber M. T. and Mouginis-Mark P. J., Constraints on magma chamber depth of the Olympus Mons volcano, in *MEVTV Workshop on the evolution of magma bodies on Mars*, LPI Technical Report **90-04**, 58–9, 1990.

Venus (Chapters 7 and 8)

Books

Barkusov, V. L., ed., *Venus Geology, Geochemistry and Geophysics. Research Results from the USSR*, University of Arizona Press, 1992.

Cattermole, P., *Venus, the Geological Story*, The Johns Hopkins University Press, 1994.

Articles

Aubele J. C. and **Slyuta E. N.**, Small domes on Venus: characteristics and origin, *Earth, Moon, and Planets, Special Venus Issue*, **50/51**, 493–532, 1990.

Campbell D. B., *et al.*, Venus: volcanism and rift formation in Beta Regio, *Science*, **226**, 167–9, 1984.

Crumpler L. S., **Head J. W.** and **Aubele J. C.**, Relation of major volcanic center concentration on Venus to global tectonic patterns, *Science*, **261**, 591–95, 1993.

Garvin J. B., **Head J. W.** and **Wilson L.**, Magma vesiculation and pyroclastic volcanism on Venus, *Icarus*, **52**, 365–72, 1982.

Head J. W. and **Wilson L.**, Volcanic processes and landforms on Venus: theory, predictions, and observations, *J. Geophys. Res.*, **91**, 9407–46, 1986.

Head J. W. and **Crumpler L. S.**, Divergent plate boundary characteristics and crustal spreading in Aphrodite Terra, Venus: a test of some predictions, *Earth, Moon, and Planets*, **44**, 219–31, 1989.

Head J. W. and **Crumpler L. S.**, Venus geology and tectonics: hotspot and crustal spreading models and questions for the Magellan mission, *Nature*, **346**, 525–33, 1990.

Head J. W., *et al.*, Large shield volcanoes on Venus: distribution and classification, *Lunar Planet. Sci.*, **XXIII**, 513–14, 1992.

Head J. W., *et al.*, Venus volcanism: classifications of volcanic features and structures, associations and global distribution from Magellan data, *J. Geophys. Res.*, **97**, 13153–98, 1992.

Head J. W., *et al.*, Venus volcanism: initial analysis from Magellan data, *Science*, **252**, 276–88, 1991.

Hess P. C. and **Head J. W.**, Derivation of primary magmas and melting of crustal materials on Venus, *Earth, Moon, and Planets*, **50/51**, 57–80, 1990.

Koch D. M., A spreading model for plumes on Venus, *J. Geophys. Res.*, **99**, 2035–52, 1994.

Phillips R. J. and **Hansen V. L.**, Tectonic and magmatic evolution of Venus, *Annu. Rev. Earth Planet. Sci.*, **22**, 597–654, 1994.

Saunders R. S., *et al.*, An overview of Venus geology, *Science*, **252**, 249–52, 1991.

Solomon S. C. and **Head J. W.**, Fundamental issues in the geology and geophysics of Venus, *Science*, **252**, 252–9, 1991.

Solomon S. C., *et al.*, Venus tectonics: initial analysis from Magellan, *Science*, **252**, 297–312, 1991.

Turcotte D. L., An episodic hypothesis for Venusian tectonics, *J. Geophys. Res.*, **98**, 17061–8, 1993.

Weitz C. M. and **Basilevsky A.**, Magellan observations of the Venera and Vega landing site regions, *J. Geophys. Res.*, **98**, 17069–97, 1993.

Io and the outer moons (Chapter 9)

Books and compilations

Bergstrahl J. T., **Miner E. D.** and **Matthews M. S.**, eds., *Uranus*, University of Arizona Press, Tucson, 1991.

Burns J. A. and **Matthews M. S.**, eds., *Satellites*, University of Arizona Press, 1986.

Gehrels T., and **Matthews M. S.**, eds., *Saturn*, University of Arizona Press, 1984.

Morrison D., ed., *Satellites of Jupiter*, University of Arizona Press, 1984.

Rothery D., *Satellites of the Outer Planets*, Clarendon Press, Oxford, 1992.

Articles

Carr M. H., *et al.*, Volcanic features of Io, *Nature*, **280**, 729–33, 1979.

Carr M. H., Silicate volcanism on Io, *J. Geophys. Res.*, **91**, 3521–32, 1986.

Greeley R., *et al.*, The Mauna Loa sulfur flow as an analog to secondary sulfur flows on Io, *Icarus*, **60**, 189–99, 1984.

Johnson T. V., *et al.*, Volcanic resurfacing rates and implications for volatiles on Io, *Nature*, **280**, 746–50, 1979.

McEwen A. S., *et al.*, Dynamic geophysics of Io, in *Satellite Phenomena and Rings*, 11–46, 1989.

McEwen A. S. and **Soderblom L. A.**, Two classes of volcanic plumes on Io, *Nature*, **280**, 736–8, 1983.

Morabito L. A., *et al.*, Discovery of currently active extraterrestrial volcanism, *Science*, **204**, 972, 1979.

Peale S. J., **Cassen P.** and **Reynolds R. T.**, Melting of Io by tidal dissipation, *Science*, **203**, 892–4, 1979.

Sagan C., Sulphur flows on Io, *Nature*, **280**, 750–3, 1979.

Schenk P. and **Jackson M. P. A.**, Diapirism on Triton: a record of crustal layering and instability, *Geology*, **21**, 299–302, 1993.

Smith B. A., *et al.*, The role of SO_2 in volcanism on Io, *Nature*, **280**, 738–43, 1979.

Soderblom L. A., *et al.*, Triton's geyser-like plumes: discovery and basic characterization, *Science*, **204**, 410–15, 1990.

Spencer J. R., *et al.*, Discovery of hotspots on Io using disk-resolved infrared imaging, *Nature*, **348**, 618–21, 1990.

Strom R. G., *et al.*, Volcanic eruption plumes on Io, *Nature*, **280**, 733–6, 1979.

Planetary geology (Chapter 10)

Books and compilations

Beatty, J. K., and **Chaikin, A.**, eds., *The New Solar System*, 1990.

Carr M. H., ed., *Geology of the Terrestrial Planets*, NASA SP-469, Washington D. C., 1984.

Cattermole P., *Planetary volcanism: a study of volcanic activity in the Solar system*, Ellis Horwood Ltd., Chichester, 1989.

Francis P. W., *Volcanoes: a planetary perspective*, Oxford University Press, 1993.

Greeley R., *Planetary Landscapes*, Unwin and Allen, 1985.

Hartmann W. K., *Moon and Planets*, Wadsworth Publishing, 1983.

Kivelson M. G., ed., *The Solar System: Observations and Interpretations*, Prentice Hall, Englewood Cliffs, USA, 1986.

Short N., *Planetary geology*, Prentice Hall, Englewood Cliffs, USA, 1975.

Taylor S. R., *Planetary Science: A Lunar Perspective*, Lunar and Planetary Institute, 1982.

Articles

Devine J. D., *et al.*, Estimates of sulfur and chlorine yield to the atmosphere from volcanic eruptions and potential climate effects, *J. Geophys. Res.*, **89**, 6309–25, 1984.

Rampino M. R., *et al.*, Volcanic winters, *Annu. Rev. Earth Planet. Sci.*, **16**, 73–99, 1988.

Taylor G. J., *et al.*, Asteroid differentiation: pyroclastic volcanism to magma oceans, *Meteoritics*, **28**, 34–52, 1993.

Wilhelms D. E., Mercurian volcanism questioned, *Icarus*, **28**, 551–8, 1976.

Wilson L. and **Head J. W.**, Ascent and eruption of basaltic magma on the Earth and Moon, *J. Geophys. Res.*, **86**, 2971–3001, 1981.

Wilson L. and **Head J. W.**, Comparison of volcanic eruption processes on Earth, Moon, Mars, Io, and Venus, *Nature*, **302**, 663–9, 1983.

Wood C. A., Monogenetic volcanoes of the terrestrial planets, *Proc. Lunar Planet. Sci. Conf.*, **X**, 2815–40, 1979.

Wood C. A., Calderas: a planetary perspective, *J. Geophys. Res.*, **89**, 8391–406, 1984.

Index

Page numbers in italic refer to illustrations